OPEN LIBRARY ⊙ SCIENCE

ELEMENTARY PARTICLES

OPEN LIBRARY ⊙ SCIENCE

General editor Frank Barnaby

ELEMENTARY PARTICLES

Frontiers of high energy physics

Gerald L. Wick

Illustrations by Sandra Barnaby

GEOFFREY CHAPMAN
LONDON   1972

Geoffrey Chapman Publishers
35 Red Lion Square, London WC1 4SJ

Geoffrey Chapman (Ireland) Publishers
5-7 Main Street, Blackrock, County Dublin

©1972, Gerald L. Wick

ISBN 0 225 65873 9

First published 1972

This book is set in 10 on 12pt, Times Roman and printed in Great Britain by A. Wheaton & Company, Exeter.

Contents

Preface	7
1. What is an elementary particle?	
The concept of atoms; what is a particle?; stability of a particle; composite particles; the world of quantum mechanics; the Heisenberg uncertainty principle	9
2. The discovery of elementary particles	
Electrons; protons; neutrons; antimatter; cloud chamber; identifying particle tracks; mesons; neutrinos; flood of particles; particle accelerators; particle detectors; strange particles; antiproton	20
3. The forces of nature	
Forces on the atomic level; strength and range of a force; the force propagator; parity; theta-tau puzzle; the Cobalt-60 experiment; neutrino's spin direction; pion decay; superweak force; superstrong force	43
4. Symmetries and conservation laws	
Momentum conservation; measuring a particle's mass; resonances; angular momentum conservation; spin quantum number; neutrino's spin; electric charge, baryon and lepton conservation; partial symmetries; parity; parity of the neutral pion; charge conjugation; isotopic spin and its violation; strange particles, associated production; strangeness	58
5. Classification of elementary particles	
The eightfold way; charge multiplets; supermultiplets; meson, baryon and high-spin resonances; hypercharge; quarks; dynamical structure; leptons; the photon	80
6. Select problems from modern research	
Internal structure of the proton and of the neutron; Rutherford scattering; the wavelength of particles; the proton's electromagnetic structure; virtual particles; virtual meson cloud; repulsive core; partons and bootstraps; CP violation in kaon decays; properties of kaons; preserving CPT; eta decay	100
Appendix: Suggestions for further reading	117
Glossary	119
Index	123

Preface

Mathematics is the language of physics. For this reason it is always difficult to write a non-mathematical book about physics which satisfies everyone from the interested layman to the professional physicist. The problem is particularly acute for books about branches of physics beyond our common experience. We are all familiar with heat, light and sound, or at least with some aspects of them. However, antimatter, isotopic spin and mesons (terms of elementary particle physics) are so removed from our daily lives that they can only be explained, in non-mathematical terms, with the use of models and analogies which are often incomplete. I have written a non-mathematical account of the most advanced and the most abstract physical science, namely, high-energy physics. High-energy physics and elementary particles are synonymous, since high energies are needed to probe the basic constituents of matter, the so-called elementary particles.

My aim is to provide a lively introduction to this subject which will serve both as a survey of particle physics in its own right and as an introduction to further study in the field. As the subject is so vast and so complex, I have included those parts of particle physics which I believe to be most important and to be amenable to explanation at an elementary level.

The concerns of professional physicists are different from the concerns of non-specialists attempting to learn about physics. Professionals worry about the digit in the third decimal place and about the rigour of a theoretical argument. These most important scientific pursuits are meaningless to non-specialists and to most students. They are interested primarily in basic ideas and how to apply them at a simple level. A trip into the submicroscopic world of elementary particles requires a good imagination coupled with an analytical mind (but not necessarily a thorough understanding of abstract mathematics). If the reader finds that this book turns him on, at whatever level, to this incredible world, then we have both been successful.

London   Gerald L. Wick
October 1971

1
What is an Elementary Particle?

'Divide and conquer' is an apt maxim for military strategists; it also describes the intellectual method used by natural philosophers in the West. In their attempts to understand matter and energy, scientists from the time of the ancient Greeks have tried to reduce the universe to its most fundamental components. Hence they 'divide' a complex system into simpler parts in the hope of 'conquering' their ignorance. This mental technique is the cornerstone of all Western rational thought and is not restricted solely to the sciences. Art and literature critics, and historians, for example, also use this approach. In the hands of scientists, this reductionist thinking led to the concept of elementary particles, the basic constituents of all matter.

There is good reason to believe that matter as we see it is composed of molecules. Molecules are composed of combinations of atoms representing the different atomic elements (hydrogen, carbon, nitrogen, oxygen, and so on). Atoms can be subdivided into compact nuclei with orbiting electrons. The nuclei are composed of elementary particles, such as protons, neutrons and other things known as mesons—to name a few. As all matter is composed of elementary particles, a comprehensive theory of elementary particles and their interactions should, in principle, explain all physical phenomena in the universe. Whether this statement is valid or not is currently an important issue in the scientific community. The pursuit of elementary particles has become 'big science', with multimillion-dollar research laboratories. Politicians, citizens, and even scientists engaged in fields other than particle physics are questioning the need for research in high-energy particle physics since it does absorb a large part of the nation's research budget. In this tense situation, the arguments are often overstated and become muddled.

Elementary particle physics is on our frontiers of knowledge. At one time atomic physics occupied this hallowed position. As higher and higher energies became available to researchers with the invention of modern particle accelerators, nuclear physics was on the edge of the intellectual wilderness before giving way to the high-energy particles. It is clear that if we are to gain new fundamental knowledge, it will most likely come from studies of elementary particles. Yet it is not so clear that this knowledge will explain everything in the universe. Most experiments are designed to study the interactions between two particles. Even a small protein molecule is composed of thousands of elementary particles. There is no guarantee that this collection of particles is just a scaled-up version of only two of them. The whole may not be just the sum of the parts. New kinds of interactions may appear as the density and number of the particles increase. There is no doubt, however, that as it now stands, elementary particle physics is one of the most exciting fields of scientific enquiry. It is certainly the most rigorous and sophisticated of sciences.

The concept of atoms

What happens if we take, for example, a piece of copper and try to split it up into smaller and smaller pieces? If matter is continuous, we would expect the copper block to be infinitely divisible. Furthermore, no matter how small we divide the bits, they should always have the same properties as bulk copper. This notion did not appeal to Democritus, a Greek philosopher who lived in the fifth century BC. He speculated that the universe is composed of small, indestructible particles called atoms (meaning 'indivisible' in Greek). According to Democritus, at a particular stage in the division of our copper block we should reach immutable atoms and the properties of the copper will change. This idea did not catch on until almost two thousand years later, when modern chemistry emerged from the Middle Ages. Until then the ideas of Aristotle and Plato held sway. They believed in an ideal world in which change was undesirable.

The concept of atoms persisted as heretical thought through the writings of the first-century BC Roman poet Lucretius. In his long poem called *De rerum natura* ('On the Nature of Things'), Lucretius wrote about atoms. He picked up the idea from Epicurus of Samos, a third-century BC philosopher who expanded the atomism of Democritus. The writings of neither Democritus nor Epicurus survived. When printing was invented in the fifteenth century, *De rerum natura* was one of the first classics to be printed.

Most scientists now believe that all matter is indeed composed of atoms. There is no way to compare the modern concept of atoms with that of the Greeks, as the ancient philosophers were not very precise about what they meant. Furthermore, what we call atoms are not indivisible, so they cannot be the same as the ancient Grecian atoms. When researchers learned to understand and control some of the forces of nature, they noticed that atoms could be subdivided. An atom is composed of a dense, central nucleus surrounded by orbiting electrons. Although the nucleus occupies less than one per cent of the volume of the atom, it contains more than ninety-nine per cent of its mass. The electrons, which carry negative electrical charges, are active in forming bonds with other atoms. Two or more atoms can coalesce to form larger molecules. Inorganic molecules, such as ordinary table salt or metal alloys, and organic molecules, such as the proteins which form our bodies, are bound by energies equivalent to the binding energy of an electron in an atom. This energy is relatively small. The energy released in a burning matchstick is sufficient to alter atomic bonds in its wood. Oxygen combines with carbon in the wood and escapes as the gas carbon dioxide.

The energy in an atomic bond is trivial compared with the binding energy of the nucleus. It takes energies five orders of magnitude greater than those in the atomic bond (that is 100,000 times larger) in order to split the atomic nucleus. Some nuclei disintegrate spontaneously. This phenomenon, called natural radioactivity, was discovered around the turn of the century. A radioactive atom becomes a different atom when it decays. For example, radioactive uranium eventually transmutes into lead. Studies of radioactivity led researchers to believe that the nucleus is composed of two kinds of particles—protons and neutrons. Different atomic elements were distinguished by the numbers of protons and neutrons in the nucleus and by the number of electrons whirling around the nucleus. See Figure 1.1.

When physicists developed machines capable of accelerating protons to very high energies (about a hundred times the energy needed to split nuclei) they started to probe the internal structure of protons and neutrons. With these studies plus studies of interactions between high-energy cosmic rays (particles, mostly protons, from outer space whose origin is unknown) and atoms in the atmosphere, the subject of 'elementary particle physics' was born. Basically, particle physics is a search for the most fundamental constituents of matter and for the laws which govern their behaviour. As the story goes, if this objective is fulfilled, all natural phenomena can, in principle, be explained. All matter is composed of elementary particles; therefore, if we understand the basic laws of physics which govern elementary particles, we should be able to understand larger systems. In this sense, elementary particles have assumed the role originally assigned by the Greeks to atoms.

What is a particle?

Now that we have the concept of elementary particles, it is not so easy to decide which bits of matter fall into this category. In the 1930s, physicists had a neat, clean picture. The elementary particles

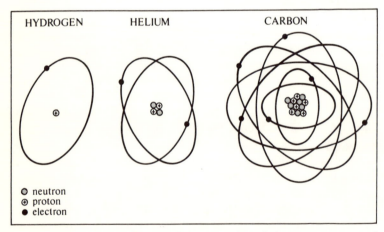

1.1. The atomic symbols for depicting the number of electrons, protons and neutrons in various atoms. The physicist's picture of the atom is more complex than these simple figures, which have little to do with the actual structure of atoms. Nevertheless it is a catchy emblem useful for advertising purposes

WHAT IS AN ELEMENTARY PARTICLE?

were electrons, protons and neutrons. However, new particles started to emerge from cosmic ray collisions and from experiments using particle accelerators. At present, the number of 'elementary' particles has exceeded a hundred, rendering the concept almost useless. As well as the three more familiar particles mentioned above, there are things called mesons, hyperons and resonances, which add considerable complications to a simple theory of matter. Out of these multitudes, some of the particles may be more elementary than others. First, however, we should have a clearer idea of what we mean by a particle and, in particular, an elementary particle.

A planetary astronomer might consider the earth as a particle. In calculations of its orbit, the earth acts as one unit and all of its mass can be treated as if it were concentrated at one point. However, we know that the earth is a composite system and cannot be very elementary. The hydrogen atom under some circumstances acts as a coherent unit, yet we know that it is composed of an electron and a proton. In a general sense, a particle is a coherent object which occupies a limited volume of space at a given time. Furthermore, it has definite physical characteristics such as mass, electric charge, and so on. Another requirement is that the particle be stable.

Stability of a particle

According to our definition, both the earth and the hydrogen atom would be particles, but the neutron would be excluded. Neutrons are as much a part of nuclei as are protons. They are perfectly stable within most nuclei, but when a neutron is alone in space, it is unstable and decays in about seventeen minutes on average into a proton, an electron and another particle called a neutrino. Since most microscopic phenomena take place during much shorter periods than seventeen minutes, we can consider the neutron as a stable particle. If we are sensible, we must relax our condition that a particle be completely stable. There are numerous particles which are more unstable than the neutron and which, as will hopefully become apparent in later chapters, should be included in our list. For example, particles known as muons decay in about 10^{-6} ($=0.000001$) seconds. Even this time is long compared to the speed at which interactions take place between particles.

According to our new criterion about stability, different configurations of the same atom would enter our list as different particles. The lowest energy configuration of an atom is stable, but because of collisions an electron can be elevated to a more energetic orbit. Within a small fraction of a second, the electron drops back down to the lowest energy orbit. Is the excited state a separate particle? The hydrogen atom has an infinite number of configurations, as does every other atom, every molecule and every nucleus. Clearly we must redefine what we mean by a particle. We cannot expect to learn much about fundamental interactions of particles when our list exceeds one million members and includes such diverse objects as planets and electrons.

Composite particles

Some of the particles are definitely composite. If we restrict membership of our list of particles to only those particles which are obviously non-composite, we can reduce our contenders to a manageable number and call them elementary particles. Yet we still run into difficulties. When one proton collides with another proton, they can create new particles, such as pions (also known as pi mesons). Does this mean that protons are composed of pions and that protons are not elementary particles? When pions collide with protons, they can create other particles known as K-mesons and lambda hyperons. The lambda hyperon decays into a proton and pion, while the K-meson decays into two or three pions. It is difficult to determine what is composed of what. One way out of this quagmire is to note that the energy of motion needed to create the new particles is comparable to the energy contained in the mass of the initial particles (using Einstein's famous equation, $E=mc^2$, where E is energy, m is mass and c is the speed of light). This situation does not apply to molecules, atoms and nuclei. With an energy expenditure which is small compared with the mass-energy of the particles involved, a molecule can be separated into its constituent atoms, an atom can be separated into its nucleus and electrons, and a nucleus can be separated into its protons and neutrons.

This hierarchy indicates that molecules are bound states of atoms, that atoms are bound states of nuclei and electrons, and that nuclei are bound states of protons and neutrons. The protons, neutrons and electrons are not in any obvious way bound states of anything

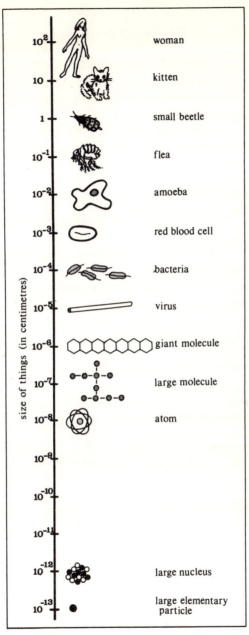

1.2. The scale of things. Atoms provide us with natural units of length which help to explain why the average man is between 5 and 6 feet tall. If all objects could be scaled up and down at will, bacteria, mice and men could all be the same size

else. They may be among the ultimate particles of nature. See Figure 1.2. In the following chapters we shall encounter other particles which are not obviously composite particles. The total number is over a hundred. Although we have improved things, the number of 'elementary' particles is still too large for satisfaction. At present there is no clear idea about which particles are the most elementary and in fact there may be no simple set of primordial particles out of which all matter is composed.

Particle physicists have made great advances in the past few years, and in later chapters I shall describe how they have managed to classify most of the elementary particles into a consistent framework. At the foundations of this framework are hypothetical particles which may indeed be the building blocks of all matter—the ultimate elementary particles. These particles, called quarks, have never been detected, despite numerous experimental searches, but the big hunt continues. Although many physicists find it philosophically pleasing to construct a hierarchy of matter with only a few elementary particles at the base, their desires may be pipe dreams. Our descriptions of nature may be much more complicated, with many particles peacefully coexisting at the most basic level. I shall discuss this point in Chapter 6.

The world of quantum mechanics

In order to appreciate fully theories of elementary particles it is necessary to delve into a physical domain completely foreign to our natural experience. The physics of small bits of matter interacting over short distances is very different from the 'classical physics' which is easily accessible to our five senses. In the first quarter of this century, the basic structure of physics underwent a profound revolution when European physicists developed for the first time a consistent theory of atomic and subatomic structure. This theory has several names—quantum mechanics, wave mechanics, or simply quantum theory; it is our best description of phenomena in the microscopic and submicroscopic world.

To the neophyte, quantum theory is so bizarre it verges on the insane. Even the physicists who gave birth to the theory had severe doubts as to its validity. In 1926, the most brilliant physicists of the day gathered around Niels Bohr in Copenhagen for a conference to discuss and review their data and ideas. One of the key members of the conference, Werner Heisenberg, recalls his memories with these uncertain words: 'I remember discussions with Bohr which went through many hours till very late at night and ended almost in despair; and when at the end of the discussion I went alone for a walk in the neighbouring park, I repeated to myself again and again the question: Can nature possibly be as absurd as it seemed to us in these atomic experiments?'

The atomic experiments to which Heisenberg referred considerably modified our concept of particles, which after 1925 became even more ambiguous than before. The experiments showed that objects ordinarily regarded as particles, such as electrons, can behave as if they were waves. Furthermore, wave phenomena, such as light, sometimes behave in a similar way to particles. This wave-particle duality is an inherent property of quantum mechanics. Electrons are neither particles nor waves, they are something else, but this something else is too difficult for our minds fully to comprehend. In the mathematical formalism of quantum mechanics, particles are described by a 'wave function'. When particles collide, their interaction can be described as the interference of two waves—as when two waves meet in a pond. When they are detected, however, they act as if they were particles. In this book I shall emphasize the particle-like properties of elementary particles.

Another aspect of the atomic experiments which troubled Heisenberg is that certain physical quantities, such as energy, only seem to be exchanged between two particles in discrete amounts, or in quanta. Consider what might happen if this quantum behaviour predominated on our scale of existence. When a person jumps into a swimming pool, for example, we know that the energy of his motion is absorbed by the water which oscillates in waves and which slightly heats up from the impact. In a quantum world, however, all of this energy would be transferred as a single bundle to another object. When one person jumps into the water, another person in the pool would absorb the energy and pop out. Of course, this picture does not make sense in our world, because the quanta are very small. What we view as a continuum is, upon closer inspection, made up of discrete intervals. In a way, the atomic view of matter is a natural quantum model. Whereas a piece of metal may look solid and continuous to our eyes, it is composed of atoms on a crystal lattice with empty space between the rows.

The Heisenberg uncertainty principle

Perhaps the most striking consequence of quantum mechanics is the 'Heisenberg uncertainty principle'. It states that there are limits to our powers of observation; in particular, both the velocity and position of a particle cannot be simultaneously measured to a high precision. This assertion seems absurd when we consider a celestial object such as a planet, whose motion and position in space we can certainly measure. With atomic particles, however, the situation is different. The best method for tracking a planet is observation of the light which reflects from it. The same technique is applicable in the microscopic realm, but the energy of the light quanta are sufficient to disturb the trajectory of the atomic particle. If I want to know where a chair is located in a room, I do not bounce cannon balls off the chair and hope to learn something when the cannon balls ricochet back. No measuring apparatus is so delicate that it does not disturb an atomic system.

This element of uncertainty, which was built into physics at the most basic level, was disturbing to many physicists and philosophers. Until his death, Einstein never liked quantum mechanics because, as he said, 'God does not play dice with the universe'. According to the new description of nature, events are governed by chance rather than predestination. Consequently the same event needs to be measured many times before the 'most probable' results can be determined.

In particle physics most of our information comes from scattering experiments where two particles collide and the initial and final states are measured. With a given initial state, the final state is not always the same. If enough collisions are recorded, however, a recognizable pattern begins to emerge. When an experimental particle physicist says that his statistics are good, he means that his data reveal a consistent pattern. The interaction of a negatively charged pion with a proton may illustrate this point.

Negative pions can be created with a particle accelerator and then directed into a target composed of hydrogen atoms, which are composed of protons and electrons. On some occasions the negative pion will bounce off the proton, rather like when two billiard balls collide. In other cases, the negative pion will exchange its charge with the

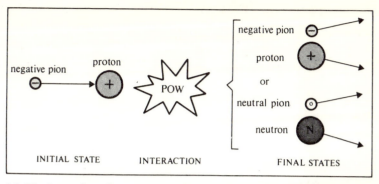

1.3. The interaction of a proton and a negative pion can lead to two different final states. Because of the random nature of the laws governing particles (quantum mechanics), it is not possible to predict the outcome of a single interaction

proton, and the final product will be a neutrally charged pion and a neutron. See Figure 1.3. In order to determine how often each final state occurs, the experimenter needs to take hundreds of measurements at each pion energy and at every possible angle, but he will never be able to predict with certainty the outcome of a single collision.

Today quantum mechanics is the basis of chemistry, of the science of materials (various solids, liquids and gases), of molecular biology, and at a more rudimentary level, of atoms, nuclei and elementary particles. Even an introduction to quantum mechanics would more than double the length of this book. I have therefore given the barest outline of some of its properties. In the remaining chapters the reader should bear in mind that with our built-in perceptual prejudices, events in the microcosm can appear very weird.

2
The Discovery of Elementary Particles

Electrons

Although the concept of an elementary particle was developed more than two thousand years ago, the first likely candidate to fill the role was experimentally identified less than one hundred years ago. Working in the Cavendish Laboratory at Cambridge University, J. J. Thomson showed in 1897 that 'cathode rays' carrying electrical current in vacuum tubes behave as particles with negative electrical charges. These particles were called 'electrons' from the Greek word which means amber. When a piece of amber is vigorously rubbed with fur, the amber picks up a static electrical charge which was arbitrarily labelled negative, the same sign as that of the electron.

It is now known that electrons are basic constituents of all atoms and that they occupy stable orbits around the atom's highly compact, central nucleus. Light is emitted when the electrons are excited from their lowest energy state and then jump from a higher orbit to another which has lower energy. This model of the atoms was developed in 1913 by Ernest Rutherford, probably the greatest nuclear physicist of his time, and by the famous Danish physicist, Niels Bohr. Rutherford was born in New Zealand and worked in Canada and in England. According to Rutherford's experiments, the diameter of an atom is more than ten thousand times larger than that of the nucleus at its centre. This means that the nucleus occupies less than one millionth of the space enclosed by an average orbit of an electron, or, if the nucleus were the size of a tennis ball, the atomic diameter would be about one mile. The next question which occupied the minds of many physicists was: What is the nucleus made of?

Protons

As the lightest of all elements, hydrogen has held a special fascination for physical scientists. In the early nineteenth century it was suggested that all atoms contain hydrogen as the basic substance. Although this idea was abandoned, it re-emerged in a modified form when physicists in the late nineteenth century noticed that cathode rays, or electrons, formed from electrical discharges in encapsulated gases, were accompanied by 'positive rays' which travelled in the opposite direction. J. J. Thomson measured properties of these rays in the 1920s. He showed that the positive rays associated with hydrogen are the lightest and thus have the smallest mass; the masses of other positive rays are multiples of the hydrogen ones. Furthermore, earlier work suggested that positive rays are none other than ions (atoms lacking one or more electrons) of the gas in the discharge tube, and in the case of hydrogen, which has only one electron, its ions are the same as its nuclei. Somewhere along the line, the positive hydrogen ion was given the name proton, from the Greek word for 'first', and it was recognized that the proton is a fundamental constituent of all atomic nuclei.

When natural radioactivity was discovered at the turn of the century, the researchers in this field could distinguish three different types of radiation. In particular, the nuclei of unstable isotopes could be transformed into isotopes of different elements with the emission of so-called alpha (α) particles, and beta (β) particles, and a single untransmuted nucleus could emit gamma (γ) rays. In time, nuclear physicists showed that alphas are the same as helium nuclei, betas are electrons and gammas are high-energy electromagnetic radiation, or photons, similar in character to visible light. These observations led theorists to speculate that the nucleus is composed primarily of protons and electrons. Alpha decay could be due to fragmentation of a nucleus, and the gamma emission could come from transitions between different stable configurations of the same nucleus, by analogy with the emission of light from transitions between different electron states in an atom.

One prediction of this model is that the isotope of nitrogen known as nitrogen-14 is composed of 14 protons and seven electrons. As the protons are about two thousand times heavier than the electron, most of the mass of nitrogen-14 must be due to the protons; the mass of nitrogen-14 was, within the experimental certainty of the 1920s, equal to 14 times the proton's mass. The seven negatively charged electrons were needed to neutralize the positive charge of seven of the protons; nitrogen-14 has a total charge of seven positive units.

Unfortunately this simple model did not work. If it did, we could wrap up this book with the conclusion that there are only two elementary particles, the proton and the electron, of which all matter is composed. Attempts to understand how the electron could be a constituent of nuclei failed. One fundamental problem became apparent when the theory of quantum mechanics or wave mechanics was developed in the 1920s. According to this theory, which still stands as the cornerstone of modern physics, each particle has an intrinsic spin, or angular momentum, which can assume only certain discrete values. These values are related to a quantum of angular momentum. We shall encounter particle spin in more detail in later chapters, but it should suffice at present to say that protons and electrons both have spin $\frac{1}{2}$ and that their spins add or subtract to give the spin of a composite particle. Thus a nucleus composed of an odd number of particles, such as proposed for nitrogen-14 with 21 total particles (14 protons plus 7 electrons), should have an intrinsic spin which is half-integral; that is, the nuclear spin theoretically could be $\frac{1}{2}$, 3/2, 5/2, and so on up to a maximum of 21/2. See Figure 2.1. Experiments with nitrogen-14, however, revealed beyond doubt that it has a quantum mechanical spin equal to the integral unit one. This paradox could be resolved if nitrogen-14 were composed of seven protons and seven other particles which are electrically neutral, which have a mass similar to that of the proton, and which possess a spin equal to $\frac{1}{2}$. Then the nucleus could be composed of an even number of particles (fourteen) whose spin angular momentum could add up to one. Such a particle, known as the neutron, was discovered in 1932.

Neutrons

In their early work on radioactivity, Lord Rutherford and his colleagues discovered that alpha particles emitted by heavy unstable nuclei can interact with stable nuclei and transform them into other atomic elements. For example, when the alpha particles (composed of two protons and two neutrons) are directed at a target of nitrogen-14 (seven protons and seven neutrons), their interaction can produce oxygen-17 (eight protons and nine neutrons) and a proton. When the alpha particles were directed at targets made up of the light elements, boron and beryllium, one of the reaction products was a very penetrating radiation. Two years after this discovery, James Chadwick, working with Rutherford at the Cavendish Laboratory, showed in 1932 that this radiation consists of electrically neutral particles having a mass similar to that of the proton.

Electrically charged particles, such as the electron and the proton,

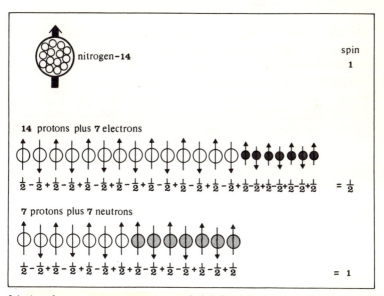

2.1. Angular momentum arguments excluded the electron as a basic constituent of nuclei. As the integral spin of nitrogen-14 necessitates that it is composed of an even number of particles with spin $\frac{1}{2}$, theoretical physicists predicted the existence of the neutron

can be easily detected. When they pass through ordinary matter, charged particles ionize the atoms along their path. These ionized atoms show up on photographic film, or they can be detected as an electrical current with simple instruments. Alternatively the original charged particles can be detected with electrical meters. None of these techniques works for neutral particles such as the neutron, which must be detected indirectly. What Chadwick did was to direct the neutrons created in the alpha-beryllium reaction into a detection chamber containing hydrogen. When a neutron struck a hydrogen atom, most of the energy was transferred to the nucleus or, in this case, the proton. The proton tracks could be accurately measured. By determining the amount of energy absorbed by the protons, Chadwick could say that the mass of the neutron is slightly greater than that of the proton.

Physicists were quite pleased with the discovery of the neutron. Atoms composed of electrons orbiting around a nucleus of protons and neutrons solved many of the problems of atomic structure. Furthermore, this model confirmed the belief that all matter is basically formed from a few elementary particles. (Incidentally, an unsophisticated explanation for beta decay and the origin of nuclear electrons is that a neutron within an unstable nucleus can spontaneously transform into a proton, an electron and a third particle called the neutrino which we shall encounter later in this chapter. When a neutron is free from its nuclear environment, it does decay to these three particles. Within stable nuclei, neutrons are inhibited from disappearing by this route.) Unfortunately, as physicists began to probe deeper into the nucleus, faith in this simple model began to waver. It finally collapsed as new particles began to proliferate in abundance.

Antimatter

The next particle to arrive on the scene conjured up wild images in many scientists' minds; it is still a favourite of science fiction writers. I am referring to antimatter or antiparticles. Each particle has an antiparticle associated with it; upon contact, an antiparticle annihilates its corresponding particle as well as itself. All that remains is either gamma radiation or mesons (see below). Four years prior to the detection of the first antiparticle, the British theorist P. A. M.

DISCOVERY OF ELEMENTARY PARTICLES

Dirac predicted their existence. In the late 1920s he developed a theory of the electron which combined Einstein's special theory of relativity with the new quantum mechanics. Dirac's equation, which is widely used by physicists today, has a mathematical solution which does not describe the electron as we know it. Rather, it describes an electron with 'negative energy'. According to Dirac's interpretation of this result, a particle with the same mass as the electron, but with positive charge, might physically exist. In 1932, C. D. Anderson of the California Institute of Technology, detected a particle in the cosmic rays impinging on the earth which had the same properties as the hypothetical Dirac particle. It is called the positron. When an electron and a positron meet they disintegrate into gamma rays.

When Dirac discovered that his equation has solutions with negative energies, he knew that if the equation was a reasonable representation of nature it would be difficult for electrons to exist. As shown in the Figure 2.2, there is an infinite number of possible states in the negative energy range; a 'real' electron will always seek its lowest energy state—similar to a ball rolling downhill. All of the negative energy states therefore need to be completely filled with electrons to prevent positive energy electrons from cascading out of the real world. According to this picture, the universe is filled with an 'infinite sea of negative energy electrons'.

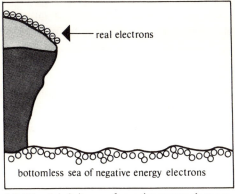

2.2. Dirac's infinite sea of negative energy electrons. Real electrons cannot drop into the sea and disappear, if the sea is completely full of negative energy electrons. If an electron is boosted out of the sea, it leaves a bubble behind which is the positron, or anti-electron

If one of the electrons in the sea receives sufficient energy, from some external source, to lift it out of the sea into the positive energy states, a 'bubble' or hole will be left behind. It is this hole which behaves as a positron. It is well known to physicists that an energetic gamma ray can create an electron-positron pair when it passes by a heavy nucleus. This phenomenon is called 'pair production' (the nucleus is basically a spectator needed to conserve momentum—see Chapter 4). Many physicists are not pleased with a universe filled with an unseen, unfelt negative-energy electron sea. Other equally bizarre interpretations of Dirac's equations have been put forward, but this one is the simplest to understand.

Cloud chamber

After Anderson's discovery of the positron, several more new particles were detected in the cosmic rays. Primary cosmic rays, which consist mostly of high-energy protons, originate from an unknown source in outer space. They interact with the earth's atmosphere and create numerous secondary particles, of which the positron is one. Several techniques have been developed to study particle interactions. Anderson used an apparatus known as a cloud chamber developed by C. T. R. Wilson in 1912 at Cambridge's Cavendish Laboratory. When charged particles pass through a gas, they ionize it. Wilson observed that in a super-saturated gas, the gas molecules will condense on the ions and form a track which can be photographed. Wilson used a mixture of air and water vapour which became super-saturated when the volume of the chamber was suddenly increased.

Charged particles can also leave tracks in photographic emulsion. Although not as common as they once were, emulsions are still useful for some types of experiments.

Identifying particle tracks

Once the tracks are obtained, it is necessary to interpret them. Physicists employ several simple tricks to distinguish the tracks left by different particles. In some cases, however, the interpretation is not so straightforward. When charged particles pass through a magnetic field, their trajectory becomes curved. The radius of the curvature depends upon the magnitude of the particle's mass and charge, and upon its velocity. The direction of the curvature depends upon whether the sign of the charge is positive or negative. Figure 2.3 illustrates the cloud chamber track which Anderson identified as the positron. The curvature of the track in the superimposed magnetic field changes as the particle passes through the lead plate. The amount of change depends on the amount of energy lost to the lead atoms by the particle. At a certain energy an electron track will exhibit a characteristic change of curvature. Anderson saw a track

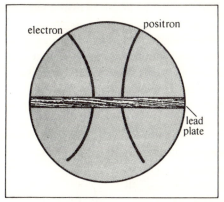

2.3. Representation of a cloud chamber photograph showing the discovery of the positron. The positron loses the same amount of energy as an electron does when it passes through the lead plate. The different direction of curvature in the magnetic field means that the positron has opposite electric charge to that of the electron

which looked exactly like an electron track, except that it was curved in the opposite direction. His interpretation, which has been verified, was that he had seen a positively charged electron, or a positron.

Another technique for distinguishing different particles depends on the number of ions they produce along their path. In a photographic emulsion, for example, the density of the track, or, crudely speaking, the thickness of the track, is used to identify the incident particle and any secondary particles.

Using both emulsions and cloud chambers, cosmic-ray researchers discovered in the 1930s and 1940s new particles with masses between those of the proton and of the electron. These particles were collectively called 'mesons', from the Greek word for 'middle'. As we shall shortly see, however, one of the original mesons was inappropriately named, for it has properties very different from the rest of the mesons.

Mesons

The meson story started with Hideki Yukawa, the great Japanese theoretical physicist. During his investigations of what binds the protons and neutrons in the nucleus, Yukawa suggested in 1935 that the nuclear force is carried by a particle about two hundred times more massive than the electron. Yukawa's idea is analogous to the theoretical situation of the electromagnetic force. The electric interaction between charged particles can be described as the exchange of a quantum (or unit of energy) of the electromagnetic field. This bundle of energy sometimes behaves as a particle and is called the 'photon', from the Greek for light. (Gamma rays as well as visible light are composed of photons.) Charged particles can exert forces on one another over very large distances and as a result the mediator of the electromagnetic field, or the photon, has a very small mass. This result is not obvious, but it is contained in the mathematical theory. Nuclear forces, in contrast, are only effective over very short distances. Experiments initially conducted by Lord Rutherford and others showed that the range of the nuclear force between nucleons (a generic name which includes both protons and neutrons) is less than 10^{-12} centimetres. Yukawa used this information to postulate the properties of his particle. These mesons are popularly described as the nuclear 'glue' which holds the nucleus together.

The physics community was very much excited in 1937 when particles presumed to be the Yukawa particles were detected in two separate cloud chamber experiments. C. D. Anderson, and independently J. C. Street and E. C. Stevenson of Harvard University, observed cosmic-ray particles with mass of about two hundred electron masses. Although they were as massive as the hypothetical Yukawa mesons, severe difficulties arose. In particular, the particle detected by Anderson did not interact as strongly with nuclei as was expected, and thus could not be the nuclear glue. For a decade the situation was muddled until three cosmic-ray physicists from Bristol

University in England found the proper particle. C. M. G. Lattes, G. Occhialini and C. F. Powell observed an unusual track in their special emulsions exposed to cosmic rays on a mountain top. Figure 2.4 is a copy of one of their emulsion photographs. The first part of the track is caused by a meson which is now known to possess properties similar to the Yukawa meson. It decays into another meson

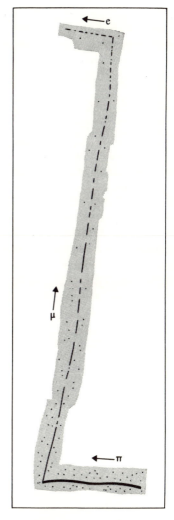

2.4. Copy of an emulsion track showing the discovery of the pion, or pi meson. It decays into a muon which then decays into an electron

which is the particle observed by Anderson ten years earlier. This particle then decays into a well-known particle—the electron. Powell and his colleagues called the primary particle a π (pi) meson and the secondary, lighter meson μ (mu) meson.

Investigations of these cosmic rays showed that the pi meson (abbreviated to pion) is about 270 times more massive than the electron. It also interacts strongly with nuclei. This property partially explains why the pi meson was never detected in cosmic rays at sea level. Pions are created at the top of the atmosphere by primary cosmic rays impinging on air molecules. The pions, however, are absorbed by the intervening atmosphere and have a small chance of reaching the ground below. Although they are seen in detectors placed on high mountains, even this is relatively rare.

On the other hand, mu mesons (now called muons, as they are very different from other mesons) readily penetrate through the atmosphere and are the main constituent of 'sea level' cosmic rays. Muons are sometimes called 'heavy electrons'. Detailed studies over the past thirty-five years have failed to reveal any differences between the electron and the muon other than properties which depend directly on their mass difference. Why the muon exists and how it fits into the scheme of things is still one of the greatest mysteries of particle physics.

When a proton and neutron scatter from each other, they supposedly exchange a charged pion between them. The pion can have either positive or negative charge. It is also possible, however, for two protons or for two neutrons to scatter via the nuclear force. In these cases, the exchanged pion cannot have electric charge. If it did, the proton would be converted into a neutron and vice versa. See Figure 2.5. To remedy this situation the British physicist N. Kemmer extended Yukawa's hypotheses about the origin of nuclear forces and suggested in 1937 that a third type of pi meson—without electric charge—exists. This neutral pion completes the explanation of the meson exchange forces between nucleons. It also helped to solve a problem relating to cosmic rays. J. Robert Oppenheimer, known in some circles as 'the father of the atomic bomb', hypothesized in 1947 that the neutral pi meson decays very rapidly into two photons when it is ejected from a nuclear reaction. The photons from this decay mode could possibly create the showers of low-energy photons and electrons observed when cosmic rays strike the atmosphere.

When man-made high-energy particle accelerators began operating in the 1950s, several searches began for the neutral pion. The experiments were not easy, since the pion has no charge and has a very short decay lifetime. At the University of California, a team of experimenters detected in emulsion two electron-positron pairs which originated from the two decay photons of the neutral pion. From these data, they could reconstruct the trajectory of the pion and the distance it travelled before decaying. This information yielded a pion lifetime of only 10^{-16} seconds. Until very recently this was the shortest lifetime to be measured directly.

Neutrinos

Meanwhile, back in the fourth decade of this century the well-known German physicist Wolfgang Pauli postulated a very peculiar

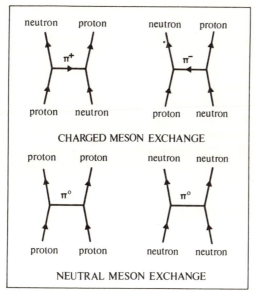

2.5. Interactions between protons and neutrons can be explained by the exchange of charged pions. A third pion, the neutral one, is needed to explain the interactions between identical nucleons. In these diagrams, called Feynman diagrams after Richard Feynman of the California Institute of Technology, time increases from the bottom of the picture to the top

particle which the Italo-American Enrico Fermi later dubbed the neutrino ('little neutral one'). Pauli proposed the existence of these chargeless, massless particles which travel at the speed of light in order to save some of physics' most cherished principles from collapse. The clues leading to the neutrino came from nuclear beta decay, which is the emission of an electron from an unstable nucleus. The problem as originally encountered by physicists in the 1930s is that conservation of energy and of momentum is violated if the electron is the only particle emitted in beta decay. Probably the most basic tenet of all physical science is that energy cannot be created or destroyed. Energy here includes mass, since there is a direct relation between mass and energy, as expressed by Einstein's famous equation $E=mc^2$, where E is energy, m is mass and c is the speed of light. Energy did seem, however, to be destroyed in beta decay. For example, in the decay of the isotope bismuth-210 (known as radium E in those days) into polonium-210, the electrons are not emitted with a well-defined energy, but have a wide range of energies extending from zero to a maximum. If energy were conserved, the electrons should always carry away the maximum amount of energy; it corresponds to the mass-energy difference between the bismuth and the polonium isotopes. Energy conservation could be saved if a particle, as yet unseen, of zero rest-mass and zero charge accompanies the electron. Pauli also predicted other properties of this peculiar particle, such as its intrinsic spin angular momentum.

Although the neutrino was a figment in Pauli's mind, the reasons for its existence were so compelling that it was easily accepted by physicists. Enrico Fermi used the concept of the neutrino in 1934 to develop a theory of beta decay which conserved energy and momentum and which reproduced the energy distribution of the emitted electrons. The basic mathematical form of Fermi's theory is still used today to describe beta decay. The theory predicts that neutrinos will interact very rarely with matter. This property makes neutrino detection very difficult, since a beam of these particles will pass virtually unattenuated through the entire diameter of the earth. Despite these poor odds, two American physicists, F. Reines and R. D. Cowan, succeeded in observing events due to neutrinos. They actually observed antineutrino events, but by the time they did their experiment in 1956 it was generally believed that every particle has an antiparticle.

Using a large flux of antineutrinos from a nuclear reactor at Savannah River, Georgia, Reines and Cowan looked for a reaction which is a variation on normal neutron decay. A neutron which is free of its nuclear environment will decay with a lifetime of about seventeen minutes into a proton, an electron and an antineutrino; this event is the simplest example of beta decay. It can be symbolically written n⟶p+e⁻+$\bar{\nu}$, where n is the neutron, p is the proton, e⁻ is the electron and $\bar{\nu}$ is the antineutrino. (ν is the symbol of the neutrino and its antiparticle is written with a bar over it. There are a few exceptions to this convention about antiparticles. One example is the positron or anti-electron, which is denoted as e⁺. It may also be written $\overline{e^+}$.) Cowan and Reines detected the product neutron and positron from the reaction: $\bar{\nu}$+p⟶n+e⁺. For all intents and purposes this reaction is the same as normal neutron decay. The arrow of time designating the direction of a reaction is not unique for microcosmic events and can be reversed for all known reactions. Furthermore, a particle when switched from one side of the equation to the other is converted to its antiparticle. The emission of a neutrino is equivalent to the absorption of an antineutrino. Thus all the following reactions are theoretically possible, although the probability for their occurrence and detection may be small:

$$n \longrightarrow p + e^- + \bar{\nu}$$
$$n + e^+ \longrightarrow p + \bar{\nu}$$
$$\bar{\nu} + p \longrightarrow n + e^+$$
$$e^- + p \longrightarrow n + \nu$$
$$e^- + \bar{n} \longrightarrow \bar{p} + \nu \quad \text{etc.}$$

In the 1960s another kind of neutrino was detected. This one is associated with the muon, while the one detected earlier is associated with the electron. It is still a complete mystery why two types of neutrino exist. If there is a chance of confusion, they are designated separately as ν_μ and ν_e respectively. Refer back to Figure 2.4, and the kinks in the tracks where the pion changes into a muon and where the muon changes into an electron can be explained by the following sequence of reactions:

$$\pi^- \longrightarrow \mu^- + \bar{\nu}_\mu \qquad \pi^+ \longrightarrow \mu^+ + \nu_\mu$$
$$\mu^- \longrightarrow e^- + \bar{\nu}_e + \nu_\mu \quad \text{or} \quad \mu^+ \longrightarrow e^+ + \nu_e + \bar{\nu}_\mu$$

As we shall see in Chapter 3, these reactions were very important in clarifying a fundamental principle of physics, namely parity conservation.

Flood of particles

In 1947 the total number of elementary particles was manageable. They included the electron, positron, photon, proton, neutron and the following two particle-antiparticle pairs—the positive and negative muons and the positive and negative pions. As well as these nine particles, five others were supported on strong theoretical grounds: the neutrino, antineutrino, neutral pion, antiproton and antineutron. Then a whole new class of particles, called strange particles, were identified following their initial detection in cosmic rays. New developments in particle accelerators also hastened the discovery of even more particles and facilitated the investigation of their properties. In eight short years particle physicists had over thirty objects which they tried to call elementary particles. There was a short lull until 1960, when the flood gates opened again with the discovery of 'resonant' particles. Their detection followed the development of a new kind of detector—the bubble chamber. Whether these extremely short-lived resonances are particles or not is a question still debated. Whatever is decided, there is no doubt that resonances will play an integral part in any theory of elementary particles. They are discussed in more detail in Chapter 5. At present there are well over one hundred 'elementary' particles, including the resonances.

The first indication that the particle situation was not so simple came in 1947 when G. D. Rochester and C. C. Butler of the University of Manchester observed tracks such as those depicted in Figure 2.6. They had exposed their cloud chamber to cosmic rays in a mountain-top laboratory and recorded these 'V particles'. In the first example two charged particles, a proton and a negative pion, were created from the decay of a neutral particle having a mass greater than that of the proton. Similar to all neutral particles, it left no track. This particle is called the lambda-zero (Λ°). In another cloud chamber photograph, a charged particle heavier than the proton was detected. This particle, known as the sigma-plus (Σ^+), and the lambda-zero are two members of a class of particles known as hyperons (from the Greek word meaning 'overmuch' or 'excessive'); all hyperons are more massive than nucleons (protons and neutrons).

Particle accelerators

Although the cosmic-ray work pointed the way, the main wealth of data was tapped with the use of particle accelerators. The first types of accelerators of interest to particle physicists were modelled after the cyclotron designed by the Californian physicist E. O. Lawrence in 1929. The early cyclotrons were not energetic enough to produce elementary particles, although they were useful for studies of the atomic nucleus. For higher energies the stability of the cyclotron is seriously undermined. The cyclotron condition requires that the particle's mass remain constant; unfortunately the mass of the proton which is being accelerated increases at high energies according to the theory of relativity. This limitation was overcome in about 1945 when V. Veksler in Moscow and independently E. M. McMillan, of the University of California, proposed a scheme for accommodating the mass change. In these accelerators known as synchrocyclotrons, the accelerated particles strike the target in bursts rather than in a continuous stream.

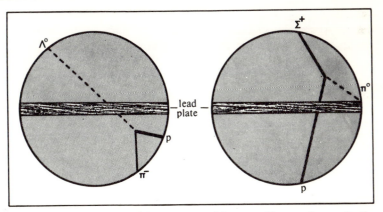

2.6. Cloud chamber tracks left by strange particles created by cosmic rays. On the left, the decay of a neutral hyperon leaves a V track composed of two charged particles. On the right a charged hyperon decays into another heavy charged particle and a light neutral particle

Physicists started probing the world of elementary particles without the aid of cosmic rays in 1948, when pions were created at the 184-inch synchrocyclotron at the University of California. This type of machine can produce protons with energies of about 300 MeV which upon colliding with a stationary target can produce ample quantities of pions with a mass of 140 MeV.[1] At about the same time, electron accelerators known as synchrotrons started producing 450 MeV electrons. This energy is sufficient to create pions. In synchrotrons the particles are accelerated around a circle. In contrast the particles in cyclotrons start at the centre of the machine and spiral to the outside as they gain energy. Linear accelerators, known as 'linacs', are also used in particle physics.

Proton synchrotrons began operation in the U.S.A. in 1955. The 3 GeV 'Cosmotron' at Brookhaven in New York and the 6 GeV 'Bevatron' at the University of California could be used to study hyperons, other strange particles, massive antiparticles and the short-lived resonances. Accelerator technology has come a long way since 1955; the U.S.A. is just completing a 500 GeV proton synchrotron (four miles in circumference) at Batavia, Illinois. The European particle physics community is starting to build a 200 GeV machine at Geneva; the Soviets held the previous lead with a 70 GeV accelerator at Serpukhov. The main limitation on the size of these machines seems to be money. They are tremendously expensive and require funding from national governments in order to be built. Since particle physics has no immediate military, technological or social spin-off, we can expect the pace to slow down.

---

[1]The electron volt (eV), a unit of energy, is often used by physicists to express the energy of motion which an accelerated particle possesses; it is also used as a measure of the rest-mass of particles from the equivalence of energy and mass as expressed by Einstein's equation $E=mc^2$. One electron volt is the amount of energy obtained by a particle, with electric charge equal to that of the electron, after it has been accelerated through an electric potential equal to one volt. One MeV (million electron volts) is equal to 10^6 eV. One GeV (giga electron volt) is equal to 10^9 eV or 1000 MeV. This symbol is preferred to BeV (billion electron volts), used in the United States, because billion means a million million in Great Britain while it means a thousand million in the U.S.A.

The mass of the electron, which has been measured as $9 \cdot 1 \times 10^{-28}$ grammes, when converted to the electron volt system becomes 0·511 MeV. The latest measurements indicate that the pion is about 270 times more massive than the electron, with a mass of about 140 MeV.

Particle detectors

The development of accelerators was accompanied by an equally impressive improvement of particle detectors and of techniques for data analysis. Most modern experiments detect particle tracks in either scintillation counters, bubble chambers or spark chambers. When charged particles pass through scintillating material, they produce a flash of light which can be detected by photoreceptors and analysed with high-speed electronics. Several scintillators can be used to define the trajectory of a particle. Often scintillators are used to trigger the operation of a bubble chamber, a device similar to a cloud chamber. Most bubble chambers contain liquid hydrogen at temperatures of about four degrees above absolute zero. The operators keep the hydrogen under compression until they want a picture. The pressure can be rapidly released and before the hydrogen boils furiously, bubbles form on the ions left in the wake of a passing charged particle. The bubbles that form when you open a bottle of beer work on a similar principle, except that these bubbles coalesce on particles of dust. In 1952 Donald Glaser while at the University of Michigan invented the bubble chamber, but it did not become a research instrument of significant stature until Luis Alvarez and his colleagues at the University of California developed bubble chambers on a grand scale. One of the largest operating bubble chambers is at CERN, the European Centre for Nuclear Research. It is four metres in length. Bubble chambers are much more versatile than cloud chambers or emulsions. They can be rapidly cycled so that many pictures can be taken in a given length of time. Further, the liquid is dense and there are many more interactions in its volume than in an equivalent volume of gas. Figure 2.7 shows a typical bubble chamber event.

Spark chambers are a more recent development. They are composed of several parallel plates with high voltage applied to alternate ones. When a charged particle passes through the stack of plates, which are enclosed in a gas-filled chamber, the gas ionizes along the particle track and an electrical spark jumps across the small gap between adjacent plates. Two cameras photograph a stereo view of the sparks which follow the particle track. Spark chambers operate considerably faster than bubble chambers. This feature helps the experimenters to trigger the detector and photograph only those

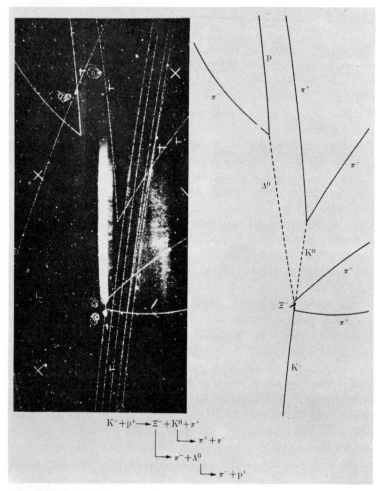

2.7. Bubble chamber photograph showing the interaction of a negative kaon with a proton. The parallel lines are tracks left by the beam of negative kaons as they pass through the liquid hydrogen. The magnetic field causes them to curve slightly to the right. At the interaction vertex the neutral kaon, produced in the reaction, leaves no track. It decays into two charged pions which leaves a V track. The negative xi hyperon decays into a neutral lambda hyperon, which leaves no track, and a negative pion. Where the lambda decays into a proton and a negative pion, another V track appears. The diagram on the right should clarify the event. (Photograph by courtesy of Rutherford High Energy Laboratory)

events in which they are interested.

The primary proton beam at the 3 GeV Cosmotron at Brookhaven was so intense it was possible to form secondary beams of pions which were then directed into a bubble chamber. Starting in 1953 this pion beam was used to create new particles from interactions with the protons in the liquid hydrogen. The properties and identity of the new particles were determined from measurements of the track curvature and length when a magnetic field was superimposed on the bubble chamber.

Strange particles

The first particles to emerge from these analyses were the K mesons (also called kaons) and the hyperons which we previously encountered in cosmic rays. They completed the nineteen particles, called the stable particles, included in Table 2.1. Yet a close look will reveal that only the photon, neutrinos, electron and proton are really stable. On a nuclear scale, however, all the remaining particles are also stable. Most nuclear events happen on a time scale of about 10^{-23} seconds. The two neutral mesons have lifetimes of about 10^{-16} seconds, which means that they exist for a length of time in which ten million events can take place. The very short-lived particles or resonances which have been excluded from the list will be discussed in Chapter 5. They exist for only 10^{-23} seconds.

Kaons and hyperons are always produced in association with each other. Two or more of them, but never one, are created in a particle reaction. This property and other unusual behaviour, such as their decay characteristics, led some theorists to suggest that they were strange. Particles which behaved in this way were assigned a new property called 'strangeness', which I shall discuss further in Chapter 4. Those particles listed in the table with non-zero entry in the column marked 'Strangeness' fall into this category.

Several more hyperons and kaons and a new meson, the eta (η), were discovered with accelerators in the few years after 1953 and they are included in the list.

Table 2.1. Stable particles. Where two numbers appear in a column, the first refers to the particle and the second to its antiparticle. If only one number appears, both the particle and its antiparticle have the same value *(Overleaf)*

Class	Name	Symbol	Anti-Particle	Mass, MeV	Spin	Parity	Isotopic Spin Total	Isotopic Spin Third Component	Baryon Number	Strangeness	Lifetime, Seconds	Principal Decay Modes
Photon	Photon	γ	—	0	1	−1	—	0	0	0	—	Stable
Lepton	Neutrino (electron)	ν_e	$\bar{\nu}_e$	0	$\tfrac{1}{2}$	—	—	—	0	—	—	Stable
	Neutrino (muon)	ν_μ	$\bar{\nu}_\mu$	0	$\tfrac{1}{2}$	—	—	—	0	—	—	Stable
	Electron	e^-	e^+	0.511	$\tfrac{1}{2}$	—	—	—	0	—	—	Stable
	Muon	μ^-	μ^+	105.6	$\tfrac{1}{2}$	—	—	—	0	—	2.2×10^{-6}	$\mu^\pm \longrightarrow e^\pm + \nu_e + \nu_\mu$
Meson	Pion	π^+	π^-	139.6	0	−1	1	1, −1	0	0	2.6×10^{-8}	$\pi^\pm \longrightarrow \mu^\pm + \nu_\mu$
		π°	—	135.0	0	−1	1	0	0	0	0.9×10^{-16}	$\pi^\circ \longrightarrow \gamma + \gamma$
	Kaon	K^+	K^-	493.8	0	−1	$\tfrac{1}{2}$	$\tfrac{1}{2}, -\tfrac{1}{2}$	0	1, −1	1.2×10^{-8}	$K^\pm \longrightarrow \pi^\pm + \pi^\circ$ $K^\pm \longrightarrow \pi^\pm + \pi^+ + \pi^-$
		K°	$\bar{K}^\circ$	497.8	0	−1	$\tfrac{1}{2}$	$\tfrac{1}{2}, -\tfrac{1}{2}$	0	1, −1	0.9×10^{-10} 5.4×10^{-8}	$K_1^\circ \longrightarrow \pi^+ + \pi^-$ $K_2^\circ \longrightarrow \pi^+ + \pi^- + \pi^\circ$
	Eta	η°	—	548.8	0	−1	0	0	0	0	2×10^{-18}	$\eta^\circ \longrightarrow \pi^+ + \pi^- + \pi^\circ$

Class	Name	Symbol	Anti-Particle	Mass, MeV	Spin	Parity	Isotopic Spin Total	Isotopic Spin Third Component	Baryon Number	Strangeness	Lifetime, Seconds	Principal Decay Modes
Baryon (Nucleon)	Proton	p^+	$\overline{p^-}$	938.2	$\frac{1}{2}$	$+1$	$\frac{1}{2}$	$\frac{1}{2}, -\frac{1}{2}$	$1, -1$	0	—	Stable
	Neutron	n	$\overline{n}$	939.5	$\frac{1}{2}$	$+1$	$\frac{1}{2}$	$-\frac{1}{2}, +\frac{1}{2}$	$1, -1$	0	0.9×10^3	$n \longrightarrow p^+ + e^- + \overline{\nu}_e$
(Hyperon)	Lambda	Λ°	$\overline{\Lambda^\circ}$	1115.6	$\frac{1}{2}$	$+1$	0	0	$1, -1$	$-1, +1$	2.5×10^{-10}	$\Lambda^\circ \longrightarrow p^+ + \pi^-$ $\Lambda^\circ \longrightarrow n + \pi^\circ$
	Sigma	Σ^+	$\overline{\Sigma^-}$	1189.4	$\frac{1}{2}$	$+1$	1	$1, -1$	$1, -1$	$-1, +1$	0.8×10^{-10}	$\Sigma^+ \longrightarrow p^+ + \pi^\circ$ $\Sigma^+ \longrightarrow n + \pi^+$
		Σ°	$\overline{\Sigma^\circ}$	1192.5	$\frac{1}{2}$	$+1$	1	$0, 0$	$1, -1$	$-1, +1$	$<1.0 \times 10^{-14}$	$\Sigma^\circ \longrightarrow \Lambda^\circ + \gamma$
		Σ^-	$\overline{\Sigma^+}$	1197.3	$\frac{1}{2}$	$+1$	1	$-1, +1$	$1, -1$	$-1, +1$	1.5×10^{-10}	$\Sigma^- \longrightarrow n + \pi^-$
	Xi	Ξ°	$\overline{\Xi^\circ}$	1314.7	$\frac{1}{2}$	$+1$	$\frac{1}{2}$	$\frac{1}{2}, -\frac{1}{2}$	$1, -1$	$-2, +2$	3.0×10^{-10}	$\Xi^\circ \longrightarrow \Lambda^\circ + \pi^\circ$
		Ξ^-	$\overline{\Xi^+}$	1321.2	$\frac{1}{2}$	$+1$	$\frac{1}{2}$	$-\frac{1}{2}, +\frac{1}{2}$	$1, -1$	$-2, +2$	1.7×10^{-10}	$\Xi^- \longrightarrow \Lambda^\circ + \pi^-$
	Omega	Ω^-	$\overline{\Omega^+}$	1672.5	$3/2$	$+1$	0	$0, 0$	$1, -1$	$-3, +3$	1.3×10^{-10}	$\Omega^- \longrightarrow \Xi^\circ + \pi^-$ $\Omega^- \longrightarrow \Xi^- + \pi^\circ$ $\Omega^- \longrightarrow \Lambda^\circ + \overline{K^-}$

Antiproton

The story of the discovery of elementary particles would not be complete without mention of the antiproton. Until 1957, when the antiproton was observed at the 6 GeV Bevatron at the University of California, the only antiparticle to be positively identified was the positron. It was by no means universally accepted at that time that all particles have an associated antiparticle. In fact, an eminent theoretical physicist had to pay off a 500 dollar bet when Emilio Segré, Owen Chamberlain, Clyde Wiegand and Tom Ypsilantis published their evidence for the antiproton. The Bevatron was designed with this experiment in mind. Older machines could not provide sufficient energy to create this one GeV antiparticle. After bombarding the bubble chamber with high-energy protons, Segré and his co-workers recorded a track which was due to the annihilation of an antiproton-proton pair. The decay products were three or more pions. The amount of energy they carried away from the interaction vertex was equal, within the limits of the experiment, to the combined masses of the annihilated pair.

Physicists now agree with the principle that all particles have an antiparticle with complementary quantum properties, such as opposite electric charge. There are three particles in Table 2.1 for which no antiparticle is listed. They are the photon and two neutral mesons—the neutral pion and the eta. There is no distinction between particle and antiparticle since these peculiar objects are their own antiparticles.

3
The Forces of Nature

From studies of the interactions between elementary particles and between more complex bits of matter, physicists have been able to divide the forces of nature into four distinct groups. One of the forces, that due to gravity, was recognized hundreds of years ago by natural philosophers and other curious souls. Rather paradoxically, the gravitational force is the least understood. Another force, that due to electricity and magnetism, is also within the realm of common human experience. Electromagnetic forces bind electrons to nuclei forming atoms, bind atoms to form molecules and thus are responsible for all chemistry and biology. Anyone who has seen a lightning bolt or has received a shock from a metal object after walking across a thick carpet on a dry day knows something about electromagnetism. The other two forces of nature fall into the realm of nuclear interactions and are not so familiar. They are called the weak nuclear force and the strong nuclear force, or weak force and strong force for short. Typical weak interactions are nuclear beta decay and the slow decays of elementary particles. Strong forces are those that bind the nucleus together. Among other reactions they involve the exchange of pions between nucleons (protons and neutrons).

One purpose of this chapter is to explore and compare some of the properties of these forces. New experimental and theoretical evidence suggests that two unusual forces may also exist. Although these new forces, the superweak and the superstrong, may be required to explain rare phenomena, the gravitational, electromagnetic, weak and strong forces most likely play the dominant role in shaping our universe.

Perhaps a few examples will serve to show generally how the forces differ. Strong interactions occur very fast. The typical time scale for a strong event is comparable to the time needed for a particle travelling at the speed of light to cross the diameter of a proton—about 10^{-23} seconds. Pions are exchanged between nucleons in a nucleus on this time scale. Particles are created in cosmic ray collisions or during bubble chamber events on this time scale. When, however, the stable particles listed in Table 2.1 decay, their lifetimes are much longer. Another process must be in operation, namely the weak interaction. Weak processes always involve either electrons or muons and their associated neutrinos. In some cases, these particles, collectively called leptons, are directly emitted in the reaction; in other cases, they may only appear as 'virtual particles' which are exchanged between other particles as the pion is exchanged during strong interactions. The time scale for weak interactions varies considerably, as can be seen from Table 2.1. Compare the neutron's decay lifetime with that of the lambda hyperon.

When photons are involved, as in the decay of the neutral pion, the interaction is electromagnetic. Electrons on their own, without neutrinos, engage in electromagnetic interactions. For example the annihilation of an electron-positron pair into photons, or gamma rays, is electromagnetic. The photon is the quantum of the electromagnetic field and as such exchanges the electromagnetic force between particles. This force acts faster than the weak force but not as fast as the strong force.

Forces on the atomic level

On the atomic scale the relative influence of the different forces is not the same as in bulk matter. For example, gravity seems to exert a large influence in our experience of life. As far as elementary particles are concerned, however, the gravitation force is so weak it can almost be neglected. The present state of experimental physics is still too crude to measure the effect of gravity for these masses. The other forces completely overwhelm any effect due to gravity and render it unobservable. Although gravitation is the only universal force—that is, it is experienced by all particles without exception—it still has not found an accepted place in the theories of elementary particles.

The other three forces are not universal; they demarcate between different kinds of particles. The electromagnetic force can be felt only by those particles which carry electric charge. Electrically neutral particles, such as the neutron, are pretty much exempted from participation in electromagnetic interactions. The strong force divides the particles into those which come under its influences— they are called hadrons and include nucleons, mesons and hyperons —and those which escaped its effects—the so-called leptons, which exclusively include the electrons, muons and neutrinos. The weak force distinguishes another symmetry of elementary particles which I shall discuss in some detail later in this chapter. The symmetry is called 'parity' or 'mirror symmetry'. For the time being we can say that the weak force operates differently on those particles which spin clockwise, when viewed along their direction of motion, from those particles which spin counterclockwise, when viewed in the same way. The abstruse comments in this paragraph should become clearer as the chapter unfolds.

Strength of a force

One way of distinguishing the forces is by the strength of their interaction. When two particles collide, such as a proton and a pion, which can interact via all of the forces, each force has a different impact on the final state of the reaction. The strong force predominates unless it is prohibited for some physical reason. Next in line is the electromagnetic force, and finally the weak force. The gravitational force is even smaller. The characteristic strength of each force can be quantified by numbers known as coupling constants or fundamental constants. In the case of the electromagnetic force, the constant is intimately related to the magnitude of the electric charge, since the size of the charge determines how strongly two charged particles interact. Appropriately defined, the constants give the following ratios for the strengths of the strong, the electromagnetic, the weak and the gravitational forces respectively:

$15 : 1/137 : 10^{-12} : 10^{-35}$

Range of a force

From a casual look at the magnitudes of the coupling constants, one might think that the strong interactions should dominate all known phenomena. This probably would be so if the strong force did not have such a short range of influence. When two particles are separated by a distance greater than about 10^{-13} centimetres, the magnitude of the strong force drops to zero. Further, some particles, such as the electron, seem to operate as if the strong force did not exist; they are not strongly interacting particles. The electromagnetic force is very long range; theoretically a charged particle should be able to feel the electric field of another charged particle which is light years away, although the force would be extremely feeble. The effects of a positively charged particle, however, can be shielded by a negatively charged particle which is nearby. Since matter is, on average, electrically neutral, any long range effects of electromagnetism are usually cancelled locally by the equal numbers of positive and negative charges.

Of the two remaining forces, only gravitational has a long range. Weak interactions between particles occur over distances which are even shorter than the range of strong interactions. On account of the impotency of the more powerful forces over long distances, the gravitational force wins by default. As far as we know, gravity has no counterpart to the positive and negative charges of electromagnetism. All matter attracts, whereas like charges repel. There is some speculation that antimatter might gravitationally repel matter. It is extremely difficult to deny or confirm this suggestion because of the infinitesimal amounts of antimatter available on our planet. Sensitive experiments at Stanford University, California, designed to compare the gravitational properties of electrons and positrons are inconclusive so far. In any case, we would not experience antigravity on our planet, but, if it does exist, it may have repercussions for theories of how the universe evolved.

Astronomers and astrophysicists have discovered peculiar celestial objects where the gravitational forces outstrip nuclear forces and furthermore, they do it over short distances. In very dense stars, known as 'neutron stars', the nuclear reactions which usually supply stellar energy have 'burned up', and the star has started to collapse from the gravitational attraction. The protons and electrons which compose the star coalesce at these high densities into neutrons and neutrinos. Under some circumstances the gravitational collapse may continue unabated, and the star disappears into a 'black hole' in space. Some astronomers believe that they have found such black holes, but their claims are hotly disputed.

The force propagator

The range of a force, or the distance over which it is effective, is intimately connected with another property of the force field—the propagator or carrier of the force. When two charged particles scatter from each other, they exert an influence by exchanging photons, the quanta of the electromagnetic field. See Figure 3.1. In this way, the photon is the propagator of the electromagnetic force. Similarly, the pions are exchanged between strongly interacting particles. There is a mathematical relationship between the mass of the propagator and the range of its associated force. The photon has a mass of zero and the electromagnetic force is of infinite range. Since the strong force has a short range, its propagator must have a finite mass. It was this concept of force propagators that led Yukawa to predict the properties of the pion. The range of the strong force gave a reasonable prediction for the pion's mass.

By analogy with the other forces, theorists believe that the weak force also has a propagator. In contrast with the other two forces, the weak force does not have a measurable range. In particular, the predominantly weak decay of the muon (to an electron and two neutrinos) involves four particles which seem to be without structure; in other words, they are all point particles. This reaction plus other evidence indicate that the weak interaction occurs at a point, or at best, it has an extremely short range. Thus the propagator must be very massive. It is called 'the intermediate vector boson', or the 'W particle' (for weak). See Figure 3.1. Latest estimates place its mass at about four proton masses, although there is some speculation that it may be twenty proton masses or higher.

So far experimental searches for the W particle have been fruitless. The best way to observe it artificially is from neutrino reactions: high energy neutrinos produced on an accelerator interact with protons in a liquid hydrogen bubble chamber. Experiments of this type are under way at the newest facilities in the U.S.A. and in

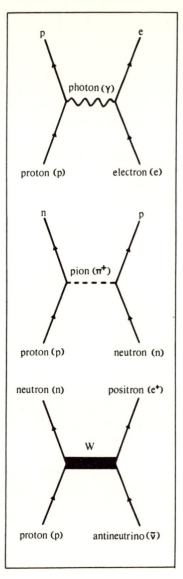

3.1. Feynman diagrams showing how electromagnetic forces are mediated by photons, how strong forces are mediated by pions and how the weak forces are mediated by the hypothetical W particle

Europe. If, however, the mass of the W is near the upper end of the theoretical range, even the newest particle accelerators will not have sufficient energy to produce it. In that case, high-energy cosmic rays may be the only source of W particles. The theoretical arguments for the W are compelling and even if it is not discovered, it will most likely be a viable particle for some time.

The gravitational field should have a propagator, called the graviton, which has zero rest mass. Since no one even knows how to treat the gravitational field using quantum theory, research into the graviton has been minimal.

Parity

One of the most intriguing chapters in the history of physics has to do with a particle symmetry called 'parity' and its relation to the forces of nature. In simple terms, parity is the 'mirror' or 'right-left' symmetry of particles and their reactions. There are many objects in our physical world which exhibit spatial symmetries. The external form of the human body is one example; on the atomic scale these examples multiply. At one time physicists thought that all of the laws of physics perfectly conserved parity; that is, they showed a complete symmetry between right and left. If a certain interaction was possible, then so was its exact mirror image. The physics community was shaken in 1956 when the Sino-Americans C. N. Yang and T. D. Lee theoretically concluded that parity conservation is violated in weak interactions. Within a few months, their calculations were experimentally verified.

Most particles have what is called intrinsic parity. Crudely this means that they are represented by a mathematical function (wave function) which is either symmetric or antisymmetric when it is reflected through a plane or a point. Figure 3.2 gives examples of both symmetric and antisymmetric functions. Those particles with symmetric wave function are said to have even ($+1$) parity; they include nucleons (protons and neutrons) and hyperons. Those particles with antisymmetric wave functions have odd (-1) parity; they include all the stable mesons and the photon. The parity of two interacting particles is equal to the product of their individual parities. In a particle reaction parity is usually unchanged by the event. For example, the reaction $\pi^- + p \longrightarrow n$ is forbidden, since the parity on the left is (−1) times (+1) equals (−1), whereas the parity of the neutron is (+1). A proper reaction would be $\pi^- + p \longrightarrow \pi^0 + n$ where the parity on both sides of the equation is (−1). Intrinsic parity, however, is not the whole story. Two interacting particles also have a relative motion whose configuration is either symmetric or antisymmetric. This relative motion depends on the 'orbital angular momentum'. The orbital angular momentum increases as the particles' energies increase. When it is zero, the parity is even (+1). This relative parity jumps back and forth between even and odd, depending on the number of quanta of angular momentum which are present in the reaction.

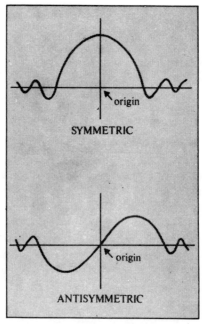

3.2. The reflection properties of a mathematical function determine its parity properties. Symmetric functions have even parity and antisymmetric functions have odd parity. All strongly interacting particles can be described by wave functions having definite parity

Theta-tau puzzle

After this short diversion we can return to Lee and Yang. They were working on the so-called theta-tau (θ,τ) puzzle when they proposed that parity conservation is not universal. During the mid-1950s experimenters studying the properties of heavy mesons identified two particles, the θ and the τ, which seemed identical in every respect except for their decay modes, which were weak interaction decays. The two particles had the same mass, charge, spin, production cross-section (probability of being created in a reaction), and so on. Both particles were unstable: the θ decayed into two pions while the τ decayed into three pions:

$$\theta^+ \longrightarrow \pi^+ + \pi^\circ$$
$$\tau^+ \longrightarrow \pi^+ + \pi^+ + \pi^-$$

As I mentioned in the last paragraph, pions have odd intrinsic parity. Thus the θ has even parity, $(-1)\times(-1)$, and in the absence of orbital angular momentum (detailed analysis confirmed this) the τ has odd parity, $(-1)\times(-1)\times(-1)$.

After examining all of the pertinent information, Lee and Yang saw that there was no evidence that parity is conserved in weak interactions. They suggested that the theta and tau were the same particle which could decay via the weak interaction in two different ways. Furthermore, since the particle could have only one intrinsic parity, one of the decay modes violates parity conservation. This particle is now known as a K meson or kaon, and it has an odd parity. It is the two pion decay (theta mode) which violates parity conservation; surprisingly this decay mode is more common than the other one.

The cobalt-60 experiment

Following their suggestion, a colleague of Lee and Yang at Columbia University, New York, verified that weak interactions violate parity symmetry. C. S. Wu joined a group of physicists at the National Bureau of Standards and measured the spatial distribution of electrons emitted by the beta decay of radioactive nuclei of cobalt-60. The reaction is $^{60}Co \longrightarrow {}^{60}Ni + e^- + \bar{\nu}$, where Ni is nickel. All particles and nuclei have an intrinsic spin angular momentum; that is, they spin around an axis through their centre. In a magnetic field the spin axes line up along the lines of magnetic force, and the particles can be considered as spinning either clockwise or counter-clockwise. If in a large collection of particles one spin state predominates over the other, the sample is said to be polarized. Wu and her collaborators polarized a sample of cobalt-60 by cooling it to low temperatures in a magnetic field. They observed an asymmetry in the number of electrons emitted along the magnetic field direction and in the opposite direction. Figure 3.3 should help to explain why this result indicates that parity conservation is violated.

As viewed along the direction of the magnetic field, the cobalt nuclei have clockwise spin—represented by the thick arrow parallel to the magnetic field direction. When the nuclei are reflected in the mirror (equivalent to the parity operation) they still spin clockwise. (The reader can demonstrate this by placing a pencil between himself and a mirror and rotating the pencil clockwise. The mirror image of the pencil will also appear to rotate in a clockwise direction from the same point of view.) An electron which travels up in the 'real' world, however, will travel down in the mirror world. As Figure 3.3 shows, if fifty per cent of the electrons were emitted along the cobalt's spin direction and fifty per cent were emitted in the opposite direction, the two worlds would have identical properties. Wu and her co-workers, however, noticed that most of the electrons were emitted in a direction opposite to that of the cobalt's spin. In the mirror world, the majority of electrons are emitted along the spin direction. Thus we have a contradiction. Here is a real event which changes when reflected by a mirror. Parity conservation must break down.

Neutrinos' spin direction

Neutrinos hold the key to parity violation in weak interactions. Careful analysis of the beta decay of cobalt-60 nuclei as well as other experiments reveals that all neutrinos are 'left-handed' and that all antineutrinos are 'right-handed'. What I mean is that the relation of a neutrino's spin to its direction of motion is similar to the action of a left-handed screw. (This description caused one famous theoretical physicist to remark that 'God is left-handed!') If a neutrino is travelling towards you, its spin is in a clockwise direction. Con-

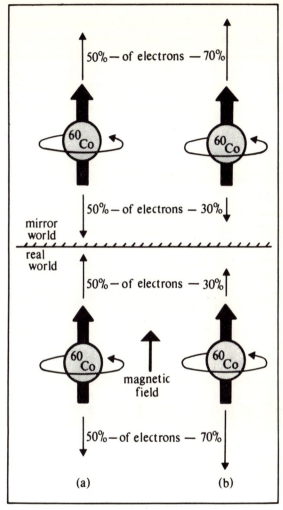

3.3. Schematic drawing of the famous cobalt-60 experiment. If 50 per cent of the electrons were emitted along the nuclear spin direction, the mirror world would be indistinguishable. An asymmetry in the electron's spatial distribution leads to a different event in the mirror world. This condition violates parity conservation

versely an approaching antineutrino is always spinning counterclockwise. As Figure 3.4 shows, neutrinos and antineutrinos are mirror images of each other; behind the mirror each disappears and reappears in a different role. Thus whenever a neutrino takes part in a reaction, parity is automatically violated; all weak interactions involve neutrinos either directly or indirectly. The spatial asymmetry of the electrons emitted from cobalt-60 is a direct consequence of the accompanying right-handed antineutrino. Of the three dominant forces among elementary particles, neutrinos seem to take notice only of the weak force. They do not participate in either electromagnetic or strong interactions. These two forces conserve parity, as far as we know.

The correlation between a neutrino's spin and its velocity means that the neutrino has a well defined 'helicity'. A corkscrew also has a well-defined helicity. The helicity of right-handed antineutrinos is defined to be positive, while the left-handed neutrinos have negative helicity. Other particles may have a definite helicity in some circumstances, but it depends on the details of the reaction and not on the inherent properties of the particle. The definite helicity of the

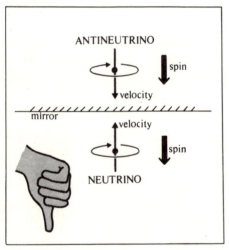

3.4. On account of the correlation between their spin and their direction of travel, neutrinos are left-handed. Upon reflection they become antineutrinos, which are right-handed

neutrino theoretically allows us to communicate our concepts of right and left to intelligent beings on other planets. Before the discovery of parity violation, philosophers and scientists both were concerned about how we would do it. Now we can give them a description of the cobalt-60 experiment and tell them to observe the antineutrino. By wrapping their right hand around the antineutrino so that their fingers follow the direction of the spin, their extended thumb should be pointing along the direction of travel. For a neutrino, the thumb of the right hand would be pointing opposite to the travel direction. See Figure 3.4. Similarly, with the left hand on the antineutrino, the thumb would point opposite to the velocity. Thus our extraterrestrial friends can discern their right from their left as we can. But there is one snag. If our comrades in outer space are made of antimatter, their experimental results will differ from ours. In particular an anticobalt-60 nucleus will emit a neutrino and a positron during beta decay. Under these circumstances, even an intelligent extraterrestrial being would confuse his left hand with his right. If, therefore, the reader ever has the opportunity to meet a cosmic pen-pal, he should beware if the friend extends his left hand in greeting—he may be made of antimatter. Very recent experiments on the eta meson may provide a way around this intrigue. They are described in Chapter 6.

Pion decay

A few weeks after the cobalt-60 experiment, parity violation was observed in another weak decay, namely:

$$\pi^+ \longrightarrow \mu^+ + \nu_\mu$$

and

$$\mu^+ \longrightarrow e^+ + \nu_e + \bar{\nu}_\mu$$

Two groups of physicists, one at Columbia University and the other at Chicago University, showed that the muon has a definite helicity following pion decay. The decay of the muon, which also violates parity, was used to determine the muon's spin direction. The muon's helicity is a consequence of the perfect helicity of the accompanying neutrino.

The forces of nature are distinguished by physical symmetries other than parity. For example there is a symmetry called isotopic spin which is violated by the electromagnetic interaction but not by the strong interaction. Isotopic spin, which we shall encounter in the next chapter, is disrupted by the presence of charged particles. Another quantum property called 'strangeness' is conserved by the strong and the electromagnetic interactions, but is violated by the weak force. This quantum number was inspired by the strange particles and will be further discussed in the next chapter.

Superweak force

As I mentioned near the beginning of this chapter, there are two hypothetical forces—the superweak and the superstrong. The superweak force was recently proposed to avoid the violation of a symmetry which is more cherished than parity, namely time reversal symmetry. According to this symmetry, the direction of time flow is not crucial to life at the atomic level. A film clip of a particle interaction would not seem funny if it were played backwards. In fact, it would be as physically possible as the forward event. Although this concept seems ridiculous from our perception of a unique direction of time flow, all the laws of physics are based upon time reversal symmetry. The difference is that we are composed of huge, vast collections of particles and atoms. Large ensembles of particles behave differently from their constituent parts.

Actually, time reversal is only a part of a larger symmetry called CPT, where the T is time reversal, P is parity and C is charge conjugation (the conversion of a particle into its antiparticle and vice versa). As I shall discuss in more detail in Chapter 6, the most fundamental laws and assumptions of physics are based on the valid conservation of this compound symmetry. If it falls, physics will need to be restructured. Experiments on kaons have shown that CP is violated—that is, if all particles become antiparticles and if the whole reaction is reflected through a mirror, the resulting reaction can be distinguished from the original one. Note that even neutrinos do not violate CP. A neutrino becomes an antineutrino from mirror reflection, or parity, but it returns back to a neutrino after charge conjugation. Kaons ordinarily decay weakly, but there is no evidence that weak interactions violate CP.

CP definitely is violated in kaon decays, and some physicists believe that T must be violated by a compensating amount so that CPT stands intact. So far neither CP nor T violation has been observed in any interaction, weak, electromagnetic or strong, except in kaon decay. The next step is to invent another force. Thus the superweak force came into existence. Supposedly it causes the CP violation in K decay and also has a compensating T violation. The other forces are left intact. This is a fortunate solution since most physicists thought that the weak interactions were going to throw another spanner in the works. Recent sophisticated experiments with kaons support some of the abstruse predictions of a superweak force. Thus it is gaining widespread favour. The only drawback is that the superweak cannot be observed in any system other than K mesons (kaons).

Superstrong force

In their search for the primordial particles, the basic building blocks of all matter, theorists have put forward all sorts of exotic creatures. The most successful one so far is the quark postulated by Murray Gell-Mann in 1964 to explain the hierarchy of particles (see chapter 5). Despite some disputed claims, quarks have not been observed. They may only be mathematical conveniences. Yet there has been speculation that quarks are so tightly bound together that we have not been able to separate them. One natural consequence of this approach is another new force—the superstrong force. This force is not universally accepted, nor is it even well-known. If quarks are discovered, the superstrong force may be given a boost. In the meantime, however, it is lying quietly in the physics archives.

4
Symmetries and Conservation Laws

When confronted with a new problem, the first thing most physicists do is look for symmetries in the physical system they are dealing with. If any symmetries exist and if they are properly identified, they provide clues to the underlying structure of the system. For example, the beautiful symmetries of crystals are due to the regular and uniform interactions of the constituent atoms. They form periodic patterns at the atomic level.

There is an intimate connection between symmetry and the so-called conserved quantities. One well-known conserved quantity is energy: it is neither created nor destroyed in any interaction (it should be remembered that mass is equivalent to energy). The corresponding symmetry in this case is time translation—that is, we should always get the same result for a given experiment independent of the time we do it. The data we collect should not depend on the day of the week we collect it, provided everything else is constant. For example, the mass of the electron, which does not depend on such quantities as the relative position of the earth with respect to the stars, should be the same value on Monday or Thursday. If this were not so, the whole basis of science would be undermined. A correct mathematical treatment of this time symmetry yields its relation to energy conservation. Thus the immutability of energy is a consequence of the properties of time.

Momentum conservation

The symmetry properties of space also lead to conserved quantities. In particular, if we move all of our experimental apparatus one foot or one mile in a straight line on the surface of the earth, it does not affect the outcome of the experiment. The decay modes of the pion should be the same in Moscow as they are in Washington. This translational uniformity of space leads to another conserved quantity, linear momentum, or as it is simply called, momentum. Momentum is defined as the product of a particle's mass times its velocity. If a particle is at rest it has no momentum, irrespective of how massive it is. If this resting particle decays into two less massive particles, momentum conservation requires that the two particles travel away in exactly opposite directions. Momentum is a vector quantity which means that momenta add or subtract, depending on their relative direction. As opposing momenta subtract, the momenta of the two final particles in our example must have the same magnitude, so that their combined momentum is zero—the same as the momentum of the initial particle prior to decay. The combined mass of the two decay products must be less than that of the original particle in order to conserve energy.

In practice, physicists have observed the following two-particle decay: $\Lambda^\circ \to p + \pi^-$. As the proton is more massive than the pion, the pion must have a greater velocity than that of the proton in the reference frame where the lambda decays at rest. The product of mass times velocity is then the same for both decay particles. The sum of their masses (938 MeV plus 140 MeV) is less than the lambda's mass (1115 MeV), otherwise the lambda could not decay into these two particles.

Measuring a particle's mass

A knowledge of energy and momentum conservation is sufficient to deduce the masses of previously unknown particles. By measuring the coordinates of the tracks left by particles in bubble chambers, the masses of new particles and of neutral particles, which leave no tracks, can be calculated. When a magnetic field is superimposed on the bubble chamber, the tracks of the particles are curved and the radius of curvature is directly proportional to the momentum of the particle leaving the track. Thus the tracks of high-energy (high velocity) particles or massive particles are not curved very much in magnetic fields. Some bubble chamber photographs reveal tracks that spiral into very small circles. See Figure 4.1. These tracks are left by either electrons or positrons, the lightest charged particles. As they lose energy through collisions in the liquid, their velocity decreases and the circles become tighter. Measurements of the track curvature, however, do not provide sufficient information to identify a particle uniquely. The density of the track also needs to be measured. The track density is a direct consequence of the number of ions produced by the particle along its path. This number depends on the velocity of the particle, and has the same form for all particles with one unit of electric charge. Track density and its curvature are usually adequate to identify a particle.

The best values for the mass of the lambda hyperon come from detailed measurements of the tracks left by the charged decay products, a proton and a pion. See Figure 2.7. The momenta and energies of the two charged particles can be calculated from the track curvature. Their momenta, however, do not add to zero, since the lambda does not have zero momentum when it decays. The lambda is created in the bubble chamber from collisions between a high-energy beam of particles, either pions or kaons, and the protons in the liquid hydrogen. The lambda gains energy from the reaction and moves with a high velocity. To remove this complicating factor, the experimenters can analyse the event in the lambda's rest frame of motion. Essentially what this means is that the equations depicting the energy and momenta of the proton and pion are transformed into a coordinate system that is moving at the same velocity as the lambda. In this moving system, the lambda appears at rest. Relatively simple manipulations of the equations now give the lambda's mass.

Resonances

Analyses of particle tracks also helped to reveal numerous short-lived particles called resonances. These particles are so short-lived that they decay before they have a chance to leave a track, but they leave an indelible impression on the energy and momenta of their decay products. The study of resonances was developed on a grand scale by Luis Alvarez and his colleagues at the University of Cali-

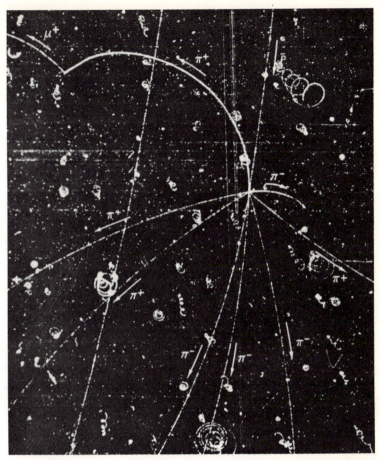

4.1. The annihilation of an antiproton in a hydrogen bubble chamber. The antiproton enters from the top right and annihilates with a proton producing eight charged pions. One of the positive pions is emitted in the backward direction with relatively low energy (as gauged by the small radius of curvature of its trajectory); it comes to rest and decays into a muon within our field of view. The tight spirals are electrons. (Courtesy of Lawrence Radiation Laboratory)

fornia. Their success depended critically on the development of large, reliable bubble chambers and of high speed computing techniques for scanning and analysing the tracks. A typical example of a resonance is the rho (ρ) meson. Its neutral member is formed in the reaction: $\pi^- + p \rightarrow \rho^\circ + n$. The rho quickly decays to two pions, so that the reaction really looks like $\pi^- + p \rightarrow \pi^+ + \pi^- + n$ on the photographs.

If the reaction proceeded without the intermediate step involving the rho, the energy would be smoothly distributed among the two pions and the neutron. If the energy and momenta of the two charged pions are measured for many such reactions and are then combined in a special way and plotted on a graph as depicted in Figure 4.2, the result should be a smooth curve. It is represented by the thick line. The presence of the rho, however, alters the distribution of the pions' energy and momenta. The initial energy of the reaction is first shared between the rho and the neutron. Then the rho imparts its energy to the two pions.

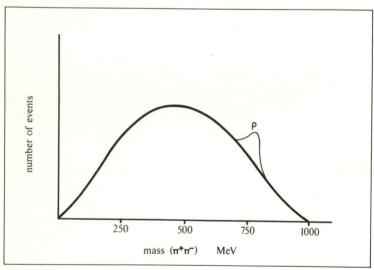

4.2. Bump on a reaction curve showing the existence of a short-lived resonance. If the momenta and energy of two pions created in a particle reaction are combined as if they were one particle, a bump, which indicates the rho meson, protrudes above the smooth background curve

In two particle reactions such as these, the possible range of energy and momenta is severely limited and we might expect to see a peak or 'bump' projecting above the three particle (two pions and a neutron) distribution. This is what happened. The peak at about 760 MeV represents the rho. The location of the peak gives the mass of the resonant particle and, by virtue of the Heisenberg uncertainty principle, the width of the peak gives its decay lifetime.

The Heisenberg uncertainty principle, which is a basic component of the quantum theory, limits the accuracy with which certain complementary quantities can be measured. If a particle is extremely stable, such as the proton, its mass can be very accurately measured. Time and energy (mass) are complementary quantities. In contrast to the proton, the resonant particles decay rapidly—in about 10^{-23} seconds—and the moment of their existence is well defined; consequently their mass is blurred out. The peak on the graph which represents the mass of a resonance is broad. For example, the rho mass is uncertain by about 125 MeV.

Angular momentum conservation

In addition to its translational uniformity, space also has a rotational uniformity. If experimental apparatus is rotated by say 90°, it should not make any difference to the experiment. This symmetry of space gives rise to another conserved quantity, angular momentum. In particle physics, angular momentum comes in two forms. When two particles rotate around each other, they have orbital angular momentum. A single particle spinning on its own axis possesses spin angular momentum, or spin. In all particle interactions the total angular momentum prior to the event equals the total angular momentum afterwards. Angular momentum conservation is also valid in larger systems; figure skaters take advantage of this fact when they draw their arms into their body in order to whirl around faster. Angular momentum is mathematically defined, for a particle, as the product of its linear momentum times its distance from the centre of rotation. When the skater pulls in his arms, their distance from the centre of his body decreases and his speed of revolution must increase in order to maintain a constant angular momentum.

Spin quantum number

In the submicroscopic world of atoms, nuclei and particles, there are restrictions imposed by quantum mechanics on the allowable values of angular momentum. It only appears as certain multiples of the basic unit, or quantum, of angular momentum. For example, all nucleons and electrons have an intrinsic spin equal to half of the basic unit. In a magnetic field, the spin axis can only point along (parallel) or opposed (antiparallel) to the direction of the magnetic field; the spin has values $+\frac{1}{2}$ and $-\frac{1}{2}$ respectively. See Figure 4.3. Spin cannot assume any of the intermediate values between these two extremes in contrast to our observations of angular momentum in the macroscopic world. Mesons have integral units of spin—that is, 0, 1, 2, and so on. Also orbital angular momentum only comes in integral units.

The number of possible quantum orientations of angular momentum relative to a fixed direction in space, such as that of a magnetic field, is twice the value of angular momentum plus one. According to this rule, the proton has $(2\times\frac{1}{2})+1$, or 2 orientations, as I have already noted. The rho meson, which has a spin of 1, has $(2\times 1)+1$, or 3, orientations—parallel, antiparallel and perpendicular to the axis in space—with value $+1$, -1, and 0 respectively. See Figure 4.3. When the angular momenta of different particles are combined, their sum must also conform to the rules. When, therefore, the spins of two protons are added, the total is either 0 or 1, depending upon their relative orientation. The total angular momentum of a proton-rho combination would be either 3/2 or $\frac{1}{2}$. If the particles rotate around each other, orbital angular momentum must be added in.

Back in chapter 2, I discussed how the neutron's existence was inferred from simple considerations of angular momentum conservation. Measurements showed that nitrogen-14 has integral spin, namely one unit. If this nucleus were composed of protons and electrons, the total number of particles would be 21 (14 protons and 7 electrons). Unfortunately for this simple idea, both protons and electrons have spin $\frac{1}{2}$ and any combination of 21 of them will have a half-integral spin. No amount of reshuffling will come up with a spin of one. Thus nitrogen-14 needs to be composed of an even number of particles possessing half-integral spin. Seven protons and seven neutrons (possessing spin $\frac{1}{2}$) fill the bill.

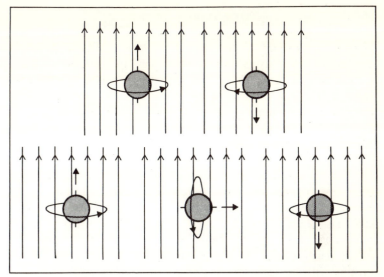

4.3. The positions of spinning particles in a magnetic field are restricted by the laws of quantum mechanics. Particles with spin $\frac{1}{2}$ (top) can align their axes with (parallel to) or against (antiparallel to) the field. Particles with spin 1 can align themselves parallel, perpendicular or antiparallel to the field

Neutrino's spin

The neutrino was postulated by Pauli in order to conserve energy and momentum in beta decay. It is also needed to conserve angular momentum in these events. Consider the beta decay of the neutron, which goes to a proton, an electron and an antineutrino. The neutron has spin $\frac{1}{2}$; so the total angular momentum after the decay must also be $\frac{1}{2}$. The total angular momentum of the proton-electron combination is either 0 or 1, and no matter how they rotate around each other, the total will never be $\frac{1}{2}$—orbital angular momentum is always integral and never $\frac{1}{2}$. Thus the neutrino with spin $\frac{1}{2}$ is needed to balance the angular momentum account book. Physicists were willing to put up with the added inconvenience of another particle in order to save their cherished conservation laws from extinction.

Orbital angular momentum is sometimes needed to balance the angular momentum on both sides of a reaction equation. Consider the rho meson which decays into two pions. As mentioned above, the rho has a spin of one. Pions possess no spin; no matter how many pions are formed, the total spin is always zero. If orbital angular momentum did not exist, it would not be possible for the spin one rho to decay exclusively into pions. However, the two pions revolve around each other so that their total angular momentum is equal to one unit. In fact, the spin of the rho was determined from the measurements of the relative motion of the two decay pions.

Electric charge conservation

In all of the millions and millions of particle reactions which have been observed, electric charge has never been created or destroyed. An implicit assumption in this statement is that the unit of electric charge is identical for all known particles. As far as we know, this is so. Yet one of the biggest enigmas of physics is the exact equivalence of the magnitude of the charge on massive particles, such as the proton, with the electron's charge. Of course, the sign is different, but this fact allows a neutral particle, such as the lambda, to decay into two charged particles of opposite sign—the proton and the negative pion. Einstein once speculated that the earth's magnetic field is due to a small, but finite, difference between the charge on the proton and that on the electron. Meticulous experiments have verified that the two charges are equivalent to better than one part in 10^{20}. This phenomenal accuracy foils Einstein's hypothesis.

The extreme stability of the electron and of the proton provides another example of how firm charge conservation is. The calculated upper limit for the lifetimes of these two particles against decay is much longer than the estimated age of the universe. If an electron does decay into two neutral particles, such as a neutrino and a photon, this process has not had an appreciable effect on the evolution of our environment. The only conservation law which inhibits this decay mode of the electron is charge conservation.

Baryon conservation

Certain types of particles never disappear unless they are annihilated by their antiparticles. Furthermore, if they are created, their antiparticle is also created. This conservation of particle number comes in two types, depending upon whether the particles are dominated by the weak or by the strong force. Furthermore, only those particles with half-integral spin are conserved. Strongly interacting particles come in two types, baryons and mesons. The baryons (Greek for 'heavyweight') include nucleons (protons and neutrons) and hyperons; they all have half-integral spin. Mesons have integral spin and include pions and kaons. Only baryons are conserved in particle interactions; mesons, as quanta and carriers of force fields, can be created or exchanged between other particles in arbitrary numbers as long as the energy is available.

The particle side of the baryons is assigned a baryon number of $+1$, while their antiparticles have baryon number -1. The sums of these numbers must balance on either side of a particle equation. For example, antiprotons are created by the following reaction: $p+p \longrightarrow p+p+p+\bar{p}$. On the left we have two protons and a total baryon number of 2. On the right the three protons add up to a baryon number of 3, but the antiproton has a value of -1 and the total is also 2. When a proton and an antiproton annihilate, they can go via the following reaction: $p+\bar{p} \longrightarrow \pi^+ + \pi^- + \pi^+ + \pi^- + \pi^°$. On the left the total baryon number is $(+1)+(-1)=0$. Pions have a baryon number equal to zero, so the total on the right is also zero. A neutron $(+1)$ cannot decay into an antiproton (-1), a positron (0) and a neutrino (0) because this event would violate baryon conservation.

The conservation of baryons is intimately connected with the stability of matter. Let us suppose that when a proton and neutron interact, they produce an extra antiproton:

$n+p \longrightarrow n+p+\bar{p}$

This event violates baryon number as it is equal to two on the left and equal to one on the right. However, if this reaction were possible, the antiproton could annihilate with the final proton and leave a neutron plus radiant energy:

$n+p+\bar{p} \longrightarrow n+\text{energy}$

The net result of this sequence of reactions is that the proton disappears:

$n+p \longrightarrow n+\text{energy}$

If this type of reaction were possible, all of the matter in the universe would have transformed into radiant energy long ago.

Lepton conservation

The lightest particles, the electron, muon and the neutrinos, do not interact strongly and are thus conserved separately from the baryons. Collectively, these particles are called leptons (Greek for 'lightweight') and the conserved quantity is the lepton number. Again, particles are assigned positive lepton number (+1) and antiparticles have negative lepton number (−1). If matters were this simple we might expect to see the following decay:

$$\mu^- \longrightarrow e^- + \gamma$$

Lepton number is +1 on both sides of the equation. This reaction has never been observed. Instead, the muon decays as follows:

$$\mu^- \longrightarrow e^- + \bar{\nu}_e + \nu_\mu$$

This observation led physicists to believe that the lepton number associated with muons and their neutrinos is conserved independently of the lepton number for electrons and their neutrinos. The muon has lepton number +1 and so does its neutrino. The electron is also +1, but the antineutrino is −1, giving a total of 0 on the right side of the equation above. The electron number is also zero on the left side. In the case of the decay of the positive muon, or antimuon, the numbers all reverse sign. We have

$$\mu^+ \longrightarrow e^+ + \nu_e + \bar{\nu}_\mu$$

and the toal muon number is −1 on both sides of the equation.

Partial symmetries

All of the conservation laws discussed so far in this chapter are obeyed by all of the forces of nature. Energy, momentum, angular momentum, electric charge, baryon number and lepton number do not change before, during or after an isolated particle interaction.

In the past few decades particle physicists have observed quantities which are partially conserved. As we saw in the last chapter, parity is conserved in strong and electromagnetic interactions, but it is not conserved in weak interactions. There are several other quantities whose conservation is violated by one or two of the forces, but not by all of them. This different behaviour of the forces of nature helps physicists to identify a particle reaction with the appropriate force.

Parity

Parity is not the first partially conserved quantity to be identified, but it is certainly the most dramatic. Prior to the theoretical work of Lee and Yang, and the experimental work of Wu, physicists thought that all particle reactions, including weak interactions, preserved their initial parity—that is, their symmetry due to mirror reflection is unchanged during the course of the particle event. All particles could be assigned an intrinsic parity quantum number which was even ($+1$) or odd (-1), depending on the symmetry of the mathematical wave-function describing the particle (see Figure 3.2). According to convention, nucleons have even parity. Armed with this information and with the knowledge that parity is conserved in strong interactions, we can determine the intrinsic parity of the other strongly interacting particles. As usual, things are not that simple; if two particles rotate around each other during an interaction, they have orbital angular momentum which modifies the reflection symmetry. If two particles rotate around each other with an even number of angular momentum quanta—that is, if the orbital angular momentum is 0, 2, 4 and so on—the total parity of the reaction is the product of the particles' intrinsic parities. If, however, they have an odd number of orbital angular momentum units (1, 3, 5, etc.), the parity reverses sign. As an example, we can determine the intrinsic parity of the neutral pion, π°.

Parity of the neutral pion

At low energy a negative pion can easily interact with a proton to create a neutron and a neutral pion:

$\pi^- + p \longrightarrow n + \pi^\circ$

The ease of the reaction indicates that there is zero orbital angular momentum leading to a conclusion that the product of the intrinsic parity of the proton with that of the negative pion is equal to the product of the parity of the neutron with that of the neutral pion. By convention both the proton and the neutron have been parity, so all pions have identical parity. However, we still have to determine whether the pion parity is even or odd.

Another reaction which proceeds relatively easily with low energy negative pions is their interaction with deuterons (nuclei of heavy hydrogen consisting of a proton bound to a neutron). The reaction is as follows: $\pi^- + d \longrightarrow n + n$, where d is a deuteron. On the left side of the equation we have the pion and the deuteron, which is composed of two even parity particles. Nuclear physics experiments have shown that the deuteron has a total even parity (the bound proton and neutron do not revolve around each other). Furthermore, the orbital angular momentum of the pion relative to the deuteron is zero because the reaction proceeds easily; extra energy would be needed to put the particles into mutual orbit. Under these circumstances, the entire parity of the pion-deuteron system is identical to the intrinsic parity of the pion. If the pion has even ($+1$) parity, its product with the deuteron's parity is even. On the other hand, an odd (-1) parity for the pion makes the product negative, or odd. If we can determine the total parity of the two neutrons, it should be the same as that of the pion.

Neutrons have even parity. Therefore the total parity of the two final state neutrons is even unless they are rotating around each other. Here we also have to consider angular momentum conservation. In the deuteron, the spins of the proton and the neutron are parallel and the total spin of the deuteron is $\frac{1}{2} + \frac{1}{2} = 1$. The pion has zero spin; the orbital angular momentum is zero; therefore, the total angular momentum is 1. In order to conserve angular momentum, the neutrons have two obvious choices (there are others, but here we should avoid making things more complicated). The neutrons can have parallel spins and zero orbital angular momentum (total angular momentum is $\frac{1}{2} + \frac{1}{2} + 0 = 1$), or they can have antiparallel spins and one unit of orbital angular momentum (total angular momentum is $\frac{1}{2} - \frac{1}{2} + 1 = 1$). The first choice is prohibited by a rule originally derived by Pauli from quantum theory. It is called the 'Pauli exclusion principle' and applies to identical particles such as two or more neutrons.

The most famous use of the Pauli exclusion principle is in atomic physics, where it places constraints on how electrons can be added to an atom. Only one electron can occupy each unique energy level, and as a consequence the electron shells, or orbits, are built up around the nucleus. Two electrons can occupy the lowest energy state, provided their spins are not in the same direction. The same situation applies to our two neutrons, but they cannot be in the lowest energy state (that is, zero orbital angular momentum) and still conserve angular momentum. The outcome of this analysis is that the neutrons have one unit of orbital angular momentum. Therefore the parity on the right side of the equation is odd, and the intrinsic parity of pions must also be odd.

The success of this logic depends upon parity conservation in strong interactions. To date, there is no evidence to contradict this conservation rule. Electromagnetic interactions also seem to conserve parity exactly, although in weak decays parity is completely violated, as we saw in the previous chapter. Hence it is meaningless to assign an intrinsic parity to leptons, the weakly interacting particles.

Charge conjugation

The symmetry between matter and antimatter brings in yet another conservation principle, 'charge conjugation'. If an interaction is invariant with respect to charge conjugation, it will make sense if all the particles are replaced by their antiparticles and vice versa. In the antimatter world, the strong interaction between a proton and a negative pion would be: $\pi^+ + \bar{p} \longrightarrow \bar{n} + \pi^\circ$ where the antiproton has a negative charge. There is no reason to believe that this reaction is not possible. Moreover, all the properties of the particle reaction and of the antiparticle reaction, such as their likelihood of happening and the angular distribution of the two final particles, should be identical. Charge conjugation seems to be invariant for both strong and electromagnetic interactions. It is violated in weak processes.

In the last chapter I discussed the famous cobalt-60 experiment which provided the first evidence that parity is violated in weak decays. In that experiment most of the electrons from the beta decay of cobalt-60 were emitted in a direction opposite to that of the nuclear spin. In the mirror world defined by parity inversion, the electrons mostly come out along the nuclear spin direction—a physically impossible situation. If, however, the nuclear particles in the parity mirror world are subjected to another reflection, charge conjugation, the situation radically alters. The cobalt-60 becomes anticobalt-60 (composed of antiprotons and antineutrons), and positrons, rather than electrons, are emitted along the nuclear spin direction. This decay is allowed and would most likely be observed in a world of antimatter. Although weak interactions violate parity (P) conservation and, independently, charge conjugation (C), they do not violate, as far as we know, the combined symmetry known as CP. The CP mirror is a symmetry salvaged for weak interactions in the wake of the fall of parity conservation. CP is most likely conserved in the other interactions as well.

Isotopic spin

When experimenters first started to explore the properties of the strong interaction, they were surprised to learn that it acted in the same way on both the proton and the neutron. This result might not be expected, since the proton has electric charge and the neutron has none, but the proton and the neutron are identical, except for their electromagnetic interaction. One might say that they are different states of the same particle. When he considered this 'charge independence of strong force', Werner Heisenberg coined a new quantum number called 'isotopic spin'. In the case of the nucleons, there are two particles and the isotopic spin is $\frac{1}{2}$; the proton has a projection on the 'isotopic spin axis' of $+\frac{1}{2}$ and the neutron has a projection of $-\frac{1}{2}$. In this formalism, the nucleons are similar to the two different projections of spin for a spin $\frac{1}{2}$ particle (when its spin was parallel to a magnetic field it was in the $+\frac{1}{2}$ state and when its spin was opposite or antiparallel to the field it was in the $-\frac{1}{2}$ state). See Figure 4.4. By all accounts, 'isotopic spin' is a gross misnomer. It is not 'spin', although the mathematics which describes it is the same as that which describes spin angular momentum. Heisenberg used the name isotopic to show that the neutron and the proton were isotopes of the same particle. Actually isobar is the correct term, since isotopes of an element all have the same electric charge. Some physicists prefer the name 'isobaric spin'.

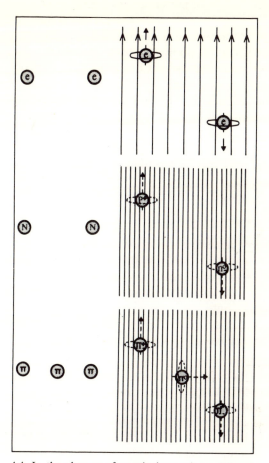

4.4. In the absence of certain interactions, different states of a particle would be indistinguishable. All electrons appear to be identical, but within a magnetic field they separate into different energy states (top). Similarly, all nucleons would be identical in the absence of the electromagnetic force. When the electromagnetic force, represented by the closely spaced parallel lines, is taken into account, isotopic spin, represented by the broken arrows, separates the nucleons into protons and neutrons (centre). Pions separate into three different charge types in the presence of electromagnetic forces (bottom)

The nucleons are called a charge doublet, since there are two particles. The pions also form a 'charge multiplet'; in this case the isotopic spin is 1 and the positive pion has projection +1, the neutral pion has projection 0, while the negative member of the multiplet has projection −1. See Figure 4.4. The projection is also called the third component of isotopic spin. In a reaction the third components of isotopic spin for the different particles add up arithmetically and the total isotopic spin adds up like vectors, according to their relative direction in 'isotopic spin space'. When, for example, a positive pion scatters from a proton, the interaction does not have much freedom in isotopic spin space:

$$\pi^+ + p \longrightarrow \pi^+ + p$$

The positive pion has third component equal to +1 and the proton's is $+\frac{1}{2}$, giving a total of +3/2. The total isotopic spin can theoretically be either 3/2 $(1+\frac{1}{2})$ or 1/2 (that is, $1-\frac{1}{2}$), but as the projection cannot be larger than the total isotopic spin, the option for total isotopic spin of $\frac{1}{2}$ is ruled out.

When a proton interacts with a negative pion the situation is quite different. The following two reactions are possible.

$$\pi^- + p \longrightarrow \pi^- + p$$
$$\pi^- + p \longrightarrow \pi^\circ + n$$

The third component of isotopic spin on the left hand side of the equations is $-\frac{1}{2}$ (−1 for the pion and $+\frac{1}{2}$ for the proton). In this case both of the total spin states are possible; it can be either 1/2 or 3/2. A total isotopic spin of 3/2 can have third components of either 3/2, $\frac{1}{2}$, $-\frac{1}{2}$, −3/2, while a total spin of $\frac{1}{2}$ only has the possibilities of $\frac{1}{2}$ and $-\frac{1}{2}$. It should be noted that in both of the reactions the third components of isotopic spin of the particles on the right hand side combine to give $-\frac{1}{2}$ in agreement with that of the initial state.

Experiments have shown that during a pion-proton interaction at energies below 300 MeV (relatively low), there is a higher probability that scattering will occur when the total isotopic spin is 3/2. Since part of the effort in the negative pion-proton interaction is absorbed by isotopic spin $\frac{1}{2}$, the probability for $\pi^- p$ scattering should be less than that for $\pi^+ p$ scattering. With the mathematics of isotopic spin it is possible to calculate that the frequency with which the three end products $\pi^+ p$, $\pi^\circ n$, $\pi^- p$ occur is respectively 9:2:1. These numbers have been verified experimentally and profoundly support the charge independence of strong interactions.

All strongly interacting particles occur in charge multiplets and can be assigned an appropriate isotopic spin. A recognition of this pattern helped theoretical physicists to see larger symmetries in the morass of particles. In the next chapter I shall mention one of the most powerful of these symmetries.

Violation of isotopic spin

As might be expected, isotopic spin is not conserved in electromagnetic interactions. The charge independence of, say, nucleon interactions is only valid if the electromagnetic force is 'turned off'. The mass difference between the proton and the neutron is probably due to the component of electromagnetic force, which interferes with these strongly interacting particles. As the mass difference is small, the component of electromagnetic force relative to that of the strong force is also small.

Even though the total isotopic spin is not conserved in electromagnetic interactions, the third component of isotopic spin is conserved. Photons can be assigned a value of isotopic spin from the decay of neutral pions:

$\pi^° \longrightarrow \gamma + \gamma$

Since the third component for the pion is zero, photons must also have a zero third component. Another example of an electromagnetic interaction is

$\Sigma^° \longrightarrow \Lambda^° + \gamma$

The sigma (Σ) is part of a charge triplet and has isotopic spin of 1, while that of the lambda (Λ) is 0. Total isotopic spin is violated by the decay, but the third component is secure.

For weak interactions, isotopic spin is meaningless. Consider the decay of the lambda:

$\Lambda^° \longrightarrow p + \pi^-$

The lambda has total isotopic spin and third component equal to zero. The pion-proton combination, as we saw, has a total isotopic spin of either 3/2 or $\frac{1}{2}$ with a third component of $-\frac{1}{2}$. This decay can be identified as a weak interaction due to its parity violation and relatively long decay time; the violation of isotopic spin and its third component add another means of identifying forces.

Antiparticles have the same total isotopic spin as their corresponding particles, but the third component is the opposite sign. Weakly interacting particles such as the electron, muon and neutrino do not have an isotopic spin quantum number, since they would not conserve it anyway.

Strange particles

From 1947, when G. D. Rochester and C. C. Butler observed the first V particles, physicists have encountered a whole new class of particles called 'strange particles'. The frequency of creation indicated that these particles were strongly interacting, but the long time scale for their decay was commensurate with weakly interacting particles. They seemed to violate a basic tenet of particle physics at that time: if a particle is created in a strong interaction, by the reversibility of reactions, it should disappear the same way. This disparity caused physicists to call these particles strange or queer.

Associated production

One of the first hints into the nature of strange particles came from A. Pais, a theoretician at the Institute for Advanced Study, Princeton, New Jersey. He noticed that the strange particles were always created in pairs in strong interactions, and, following their separation, they decayed individually through the weak interaction. He called his hypothesis 'associated production'. In a bubble chamber photograph taken at Brookhaven, New York, in 1954, a perfect example of Pais' proposal was discovered. Figure 4.5 illustrates the tracks. Two Vs turned up after the interaction of negative pion in a hydrogen chamber. Obviously two neutral particles were created at the same point because the vertices of the Vs pointed to the end of a pion track which terminated in a nuclear interaction. The masses of the neutral particles could be calculated from the curving trajectories of their decay products. The distances which the neutrals travelled before decaying gave a value for their lifetimes. The reaction which takes place at the end of the pion

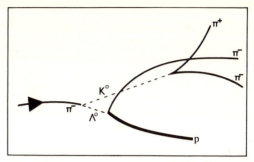

4.5. Strange particles are always produced in pairs. This phenomenon, called associated production, is depicted here by the interaction of negative pion with a proton in a hydrogen bubble chamber. The two strange particles have no electric charge and do not leave tracks

track is:

$$\pi^- + p \longrightarrow K^\circ + \Lambda^\circ$$

And the decay modes are:

$$\Lambda^\circ \longrightarrow p + \pi^-$$
$$K^\circ \longrightarrow \pi^+ + \pi^-$$

Both the lambda and the kaon decay into strongly interacting particles, but the lifetimes are very long.

Strangeness

In order to explain the strange phenomena, Murray Gell-Mann of the California Institute of Technology and, independently, K. Nishijima of Japan proposed a new quantum number called 'strangeness', S. According to their scheme some particles carried a finite strangeness, just as they possess spin, parity, or isotopic spin. In strong interactions this strangeness is conserved, while in weak interactions it is violated. All 'normal' particles have zero strangeness; in the reactions listed above, the neutral kaon was assigned strangeness equal to $+1$ and the lambda was given strangeness -1. Thus when they are created by the pion-proton reaction, strangeness is conserved (the strangeness quantum numbers add together). When the particles are separated they cannot decay into lighter particles via the strong interaction and still conserve strangeness.

The strangeness quantum number is listed along with other properties in Table 2.1 (see pp. 40-1 above). The assignment of a consistent set of strangeness numbers is relatively easy. From the initial assignment for the neutral kaon and the lambda the following reactions show how some of the other particles got their strangeness:

$\pi^+ + n \longrightarrow K^+ + \Lambda^\circ$ $S = +1$ for K^+
$\pi^- + p \longrightarrow K^\circ + \overline{K^-} + p$ $S = -1$ for $\overline{K^-}$
$\pi^+ + p \longrightarrow \overline{K^\circ} + K^+ + p$ $S = -1$ for $\overline{K^\circ}$
$\pi^- + p \longrightarrow K^+ + \Sigma^-$ $S = -1$ for Σ^-
$\pi^+ + n \longrightarrow K^\circ + \Sigma^+$ $S = -1$ for Σ^+
$\Sigma^- + p \longrightarrow n + \Sigma^\circ$ $S = -1$ for Σ°
$\pi^- + p \longrightarrow K^\circ + K^+ + \Xi^-$ $S = -2$ for Ξ^-

Certain particles with the same strangeness seem to belong to a charge multiplet similar to the nucleons or the pions. Two prominent groups are the kaon doublet (K^+, K°) with strangeness $+1$, and sigma triplet (Σ^+, Σ°, Σ^-) with strangeness -1. This observation helped theoreticians to classify particles, as I shall show in the next chapter. The strangeness of antiparticles is opposite to that of their corresponding particles.

The strangeness quantum number 'explained' why certain reactions, such as $n + n \longrightarrow \Lambda^\circ + \Lambda^\circ$ and $\pi^- + p \longrightarrow \Sigma^+ + K^-$, were never observed. These two reactions are allowed by most of the conservation laws except the new strange one. Those particles which violate strangeness when they decay do not have sufficient mass-energy to decay into other strange particles. If they did have enough mass, the decay could proceed via the strong interaction. There are strange particle resonances which are very short-lived because they have an available outlet to lighter strange particles.

Strangeness is conserved in electromagnetic interactions if the photon has a strangeness of zero. In the following electromagnetic decay where isotopic spin is violated, strangeness is conserved (see Table 2.1 on pp. 40-1 above for the quantum numbers):

$\Sigma^\circ \longrightarrow \Lambda^\circ + \gamma$

No strangeness is assigned to leptons, since weak interactions do not conserve the quantity.

Strangeness is closely related to other quantum numbers. If charge is designated by Q, strangeness by S, baryon number by B, and the third component of isotopic spin by I_3, these four quantities satisfy the following equation for any strongly interacting particle: $Q = I_3 + \frac{1}{2}(S+B)$. Because of this interdependence of these four quantum numbers, only three of the four are needed to specify completely a strong interaction. The strangeness quantum number is not necessarily new and is not even needed. However, it does help physicists to visualize what is occurring in the interactions.

Table 4.1 lists the important conserved quantities and summarizes how the three dominant forces treat them.

	Interaction		
Quantity	Strong	Electromagnetic	Weak
Energy	X	X	X
Linear momentum	X	X	X
Angular momentum (spin)	X	X	X
Electric charge	X	X	X
Baryon number	X	X	X
Parity	X	X	
Charge conjugation	X	X	
Strangeness	X	X	
Isotopic spin	X		
Third component of isotopic spin	X	X	

Table 4.1. Forces of nature and their symmetries. Each force conserves only those quantities where an X appears

5
Classification of Elementary Particles

When the symmetry of the periodic table of the chemical elements was first proposed by the great Russian chemist Mendeleev in 1869, he and his contemporaries could not explain its simple structure. How the regular progression in mass from the lightest element (hydrogen) to the heaviest naturally occurring one (uranium) related to the regular pattern of their chemical activity baffled the best minds of that time. It took Rutherford and Bohr, who used new information as well as the earlier results, to piece together a reasonable model of atomic structure which explains the periodic table. The Rutherford-Bohr atom with a compact massive nucleus having positive charge surrounded by negatively charged electrons in stable orbits reached the stage of a well-developed theory when quantum physics was developed primarily in Germany in the 1920s by Max Born, Werner Heisenberg and Erwin Schrödinger. The Frenchman Louis de Broglie also played a decisive role.

A similar situation existed four hundred years earlier in the fields of astronomy and planetary physics. Copernicus saw the symmetry when he refuted the Ptolemaic idea that the earth is the centre of the universe. The Copernican revolution radically changed man's place in nature by shifting the centre of the solar system to the sun, but it took Galileo and Newton to find the dynamical laws which govern the motions of the planets.

The eightfold way

Particle physics is still in the Mendeleevian stage of development. Only a little over a decade ago it was pre-Mendeleevian. In 1961, Murray Gell-Mann, of the California Institute of Technology, and independently, Yuval Ne'eman, an Israeli army colonel and engineer-turned-physicist, noticed a regular pattern among all the strongly interacting particles—including stable and semi-stable ones and the very short-lived ones (resonances). They worked out a scheme based on a little-known mathematical symmetry known as SU(3) and were able to predict the properties of several particles prior to their detection. The success of this model has encouraged particle physicists who are now waiting for the modern-day equivalents of Rutherford and Bohr to appear. Gell-Mann dubbed the model the 'eightfold way' after an aphorism attributed to Buddha; SU(3) symmetry is based on eight quantum numbers.

Charge multiplets

The first clue about how to classify strongly interacting particles came with the recognition of charged multiplets and the development of isotopic spin. For example, the nucleons consist of two particles, the proton and the neutron. Both of these particles have spin equal to $\frac{1}{2}$ unit, they both have even (+) intrinsic parity (the spin and parity of these particles are written with the simplified notation, $\frac{1}{2}^+$), and when strangeness was invented both nucleons were unequivocally assigned zero as their strangeness quantum number. The only difference between the two particles is their electric charge. Using the formalism for isotopic spin, which I discussed in the previous chapter, the two nucleons have the same total isotopic spin, equal to $\frac{1}{2}$, but each particle has a different projection in 'isotopic spin space'. The proton has a projection, or third component, equal to $+\frac{1}{2}$, and the neutron is $-\frac{1}{2}$. The small difference in the mass of the two particles is probably due to the electromagnetic force which does not conserve isotopic spin. It is no trivial coincidence that the mass difference between the proton and the neutron—about one part in 1000—is the same as the relative strengths of the strong and electromagnetic interaction.

There are several other examples of charge multiplets containing particles with identical parity, strangeness and spin. The pions form a triplet with isotopic spin of 1; the three members have third component of isotopic spin of $+1$ (positive pion), 0 (neutral pion) and -1 (negative pion). Their spin and parity is 0^- (zero spin and odd parity) and strangeness is zero. Among the 'stable' baryons, of which the nucleons are members, there is another charge doublet—the xi particle, or cascade particle as it is sometimes called. In this case the two members have zero electric charge (Ξ°) and negative electric charge (Ξ^-). They have spin and parity of $\frac{1}{2}^+$ and strangeness of -2. The stable baryons also include a charge triplet (isotopic spin equal to one) and a charge singlet (isotopic spin equal to zero). The positive, neutral and negative charged sigma (Σ^+, Σ°, Σ^-) comprise the triplet; their spin and parity is $\frac{1}{2}^+$ and strangeness is -1. The lambda hyperon (Λ°) is in a class by itself. The singlet also has spin and parity of $\frac{1}{2}^+$ and strangeness of -1. The quantum numbers for antiparticles are reversed whenever possible. Their total isotopic spin, spin, parity and mass, are the same as for their corresponding particles. For antiparticles, however, the following quantum numbers have opposite sign: the electric charge (third component of the isotopic spin), baryon number, and strangeness.

Baryon supermultiplet

The reader may have noticed that all the stable baryons have the same spin and parity, namely $\frac{1}{2}^+$. This identity certainly did not escape the notice of Gell-Mann and Ne'eman. They thought that all these particles might belong to a larger multiplet, or supermultiplet, which connects both different isotopic spin and different strangeness. The charged multiplets only connect particles having different values for the third component of isotopic spin. The complicated mathematics of SU(3) provided a scheme for arranging all of the baryons with spin $\frac{1}{2}$ and positive parity. By plotting strangeness on one axis and the third component of isotopic spin on the other, the particles line up in a symmetrical pattern as shown in Figure 5.1. Both the neutral sigma and the lambda line up at the same spot.

There is a deeper significance to this SU(3) arrangement than just a pretty pattern. The mass difference between the different multiplets is tied to the form of the symmetry and to the forces which give rise to it. If the electromagnetic force did not violate isotopic spin conservation, the members of a charged multiplet would all have the same mass; most likely, they would be indistinguishable or, in the parlance of quantum theory, 'degenerate'. A similar situation holds for the supermultiplet of all of the stable baryons. All of the particles in the supermultiplet would have identical mass if the symmetry were exact. However, the differences between the masses characteristic of each charge multiplet are about ten times that of the mass difference within each charge multiplet. Furthermore, the masses of the baryons are about ten times larger than the mass difference between charged multiplets. This progression indicates that the symmetry which gives rise to the supermultiplet is violated by a force which is stronger than the electromagnetic force. It is probably violated by one part of the strong force. It may be that the SU(3) symmetry is completely violated by the strong interaction. If this is so, some new force must conserve the symmetry; some theorists believe that we are in store for an even stronger force which has not yet been revealed (perhaps the superstrong force discussed in chapter 3). Experiments at the extreme-

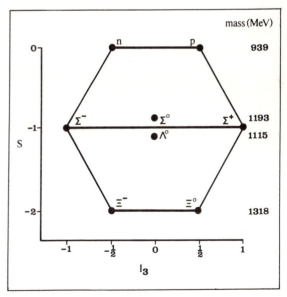

5.1. SU(3) representation of the stable baryon supermultiplet

ly high energies of cosmic rays or maybe at the modest energies of the largest man-made accelerators may prove them correct.

Violations of the SU(3) symmetry affect the masses of the multiplets within the supermultiplet. The form of the symmetry provides a rule which connects the masses of the multiplets. This rule should hold if the violation is not too violent. For stable baryons the rule is that $\frac{1}{2}$ the nucleon mass plus $\frac{1}{2}$ the xi mass should equal 3/4 the lambda mass plus $\frac{1}{4}$ the sigma mass. If the actual masses of the four multiplets are substituted into the equation, the validity of this approximate rule is surprisingly good.

Meson supermultiplet

The success of SU(3) symmetry in grouping the lightest baryons led the particle theorists to look for connections between families of mesons. At that time only a few mesons had been identified. It was 1961 and only the pion charge triplet and the kaons were known for certain. They all have zero spin and odd parity, that is 0^-. The isotopic spin of the kaons was not obvious at first, but with the invention of strangeness and with the understanding of how the different forces of nature conserve the different symmetries, kaons could be assigned an isotopic spin of $\frac{1}{2}$. On the particle side there is a doublet, the K^+ and the K°, with a corresponding doublet of antiparticles, $\overline{K^\circ}$ and $\overline{K^-}$.

The analogy with the stable baryons would be complete if there were a charge singlet with spin and parity of 0^-. The mass of this meson can be predicted from a mass rule similar to the baryon one. There are a few differences, however, between the baryon supermultiplets and the meson supermultiplets. The meson groups include both particles and antiparticles, whereas the baryon supermultiplets are wholly particles. There is a complementary baryon supermultiplet composed of antiparticles. Further, in the mass rule the mesons' masses are squared before they are inserted in the equations. With this knowledge, the mass of the last member of the meson octet could be predicted. This meson was found in late 1961 and conformed to all its anticipated properties. The eta (η) with a mass of 549 MeV completed the first meson octet as depicted in Figure 5.2. It was one of several meson resonances to be discovered in 1961.

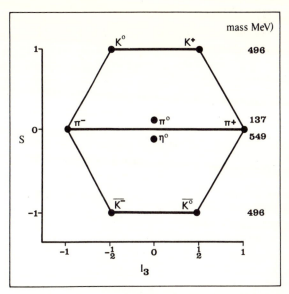

5.2. SU(3) representation of the stable meson supermultiplet

Meson resonances

As I mentioned in the previous chapter, the presence of particle resonances can be deduced from the trajectories of the final particles created in a high-energy particle collision. Consider the annihilation of a proton-antiproton pair. In some cases they can transform into five pions, two positively charged, two negatively charged and one neutral one:

$$p + \bar{p} \longrightarrow \pi^+ + \pi^+ + \pi^- + \pi^- + \pi^\circ$$

If the energy and momenta of a neutral combination of any three pions (that is, a neutral, a positive and a negative) are plotted as if they came from one particle, the eta particle (which can decay into three pions) appears as a bump in the curve. See Figure 5.3. There is another bump on the three pion curve. This one represents a meson resonance called the omega (ω). It also decays into a neutral combination of three pions but has a larger mass (784 MeV). Both the eta and the omega decay so rapidly that they do not travel far enough, even at the speed of light, to leave a measurable track in a bubble chamber. Yet their lifetimes can be determined from the width of their respective bumps. According to the Heisenberg uncertainty principle (see previous chapter) a narrow bump means a long lifetime. The omega conforms to the norms of a strongly interacting particle and decays in about 10^{-23} seconds. The eta, however, is a narrower resonance and takes as long as 10^{-18} seconds to decay. Furthermore, studies of its other decay modes have shown that the eta does not conserve isotopic spin symmetry. These two properties of the eta exclude its decay by strong interaction. But its decay nicely agrees with the characteristics of the electromagnetic interaction. Because of its relatively long lifetime, the eta joins the other stable particles and is included in Table 2.1 (see pp. 40–1 above).

Analyses of the three pion combinations can give the spins of the short-lived particles. The parity and other quantum numbers can also be extracted from the interaction. As expected, the eta has spin and parity of 0^-. In the case of the omega, however, physicists noticed that the three pions must be revolving around each other. The orbital momentum amounts to one unit. Therefore the omega has spin equal to one; its parity is negative. Other meson resonances

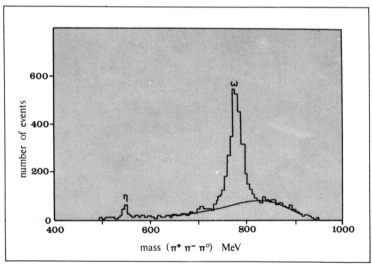

5.3. Curve of the combined mass of three pions. The two bumps represent the three-pion resonances, the eta meson and the omega meson

with spin and parity of 1^- were detected in the early 1960s. The rho (ρ) meson, which we have encountered earlier in the book, was one of them. It comes in three charge states making it a triplet. The omega is a singlet. To complete the SU(3) supermultiplet a charge doublet and its antiparticle resonances are needed. The K* (pronounced K star) resonance was actually the first meson resonance to be detected in that fruitful year 1961. It has the correct properties to fit into the meson supermultiplet with spin and parity of 1^-. Its mass and other properties also conform to the theory. It comes in positively charged and electrically neutral forms on the particle side with the appropriate antiparticles. The octet is shown in Figure 5.4.

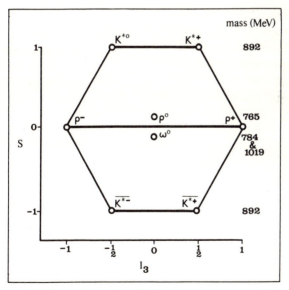

5.4. SU(3) representation of the meson resonances. The omega has two mass values because a similar ninth particle is associated with the basic octet

Following the successful classifications of meson resonances, 'bump hunting' became the prime interest of many particle physicists. Many resonances have been detected in recent years. If, however, an insufficient amount of data is available, the researchers sometimes believe that they have detected a bump in their experimental curves when none is actually present. Thus they mistake a statistical fluctuation in their data points for a resonance. To prove that his colleagues could be over-zealous, one physicist generated a false experimental curve without even doing an experiment. He then distributed the data to several peers and asked them to analyse the curve for resonances. As he expected, their results did not agree. When, however, there is a real dispute, additional data usually resolve the discrepancy.

Baryon resonances

The most widely acclaimed success of the eightfold way was in the classification of baryon resonances. The detection of the omega-minus (Ω^-: note that this is a different particle from the omega meson, which is written in the lower case, ω) in 1964 was the crowning glory of this model of elementary particles. The first baryon resonances were detected by Enrico Fermi and his collaborators at the University of Chicago in 1952. When they scattered pions on protons, they discovered that the cross-section for the reaction, that is, the probability that an interaction takes place, increased rapidly at a pion energy of about 180 MeV and decreased just as rapidly at a slightly higher energy. See Figure 5.5. In crude terms, the bump in the cross-section represents a momentary marriage of the pion and proton into a new short-lived particle. This bump is present in the curves for proton scattering with both positive and negative pions. More bumps appear for scattering with even higher energy pions, indicating more massive short-lived particles. The lowest energy bumps are called the N* (N star) to indicate that they may be excited states of the nucleon. Their mass is about 1238 MeV. Some physicists call these resonances delta ($\triangle$) particles. Measurement of all the possible pion-nucleon combinations shows that the $\triangle$ belongs to a charge multiplet with four members. The isotopic spin is 3/2 and the spin and parity are $3/2^+$.

Additional baryon resonances were discovered by Luis Alvarez and his team using the bubble chambers at the University of

California. The first strange resonance to be detected in this way was the Y* from the interaction of a lambda and a pion. This new type of resonance is a charge triplet having spin and parity of 3/2⁺ and a mass of 1385 MeV. Since the △ is a quartet, it cannot belong to an octet similar to the one for the stable baryons. Fortunately the SU(3) system allows higher order supermultiplets. The next simplest one is a decuplet with 10 members. According to the theory, the

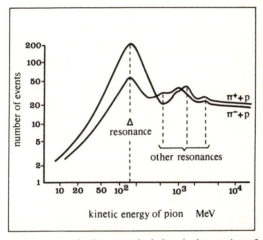

5.5 Resonances in the curve depicting the interaction of pions with protons. The lowest energy bumps correspond to the delta particles

decuplet should contain a quartet, a triplet, a doublet and a singlet. In this case, the mass rule differs from that for the octet. The mass separation between the quartet and the triplet should equal that between the triplet and doublet and should also equal that between the doublet and the singlet. From the spacing between the △ and the Y*, Gell-Mann predicted that two particle resonances (a charge doublet) with spin and parity of 3/2⁺ should appear at 1532 MeV.

Furthermore they should have a strangeness of -2 and isotopic spin $\frac{1}{2}$. And there should be a single particle with strangeness -3 and a mass of 1676. The decuplet is shown in Figure 5.6.

Alvarez and his colleagues found the doublet known as Ξ^*(xi star) which rapidly decays into a Ξ (strangeness -2) and a pion. Its mass of 1530 is remarkably close to prediction. The remaining particle should have a strangeness of -3. As there are no lighter baryons with this large strangeness, Gell-Mann and Ne'eman suggested that it cannot decay via the strong interaction and will be long-lived as a consequence. During a 1962 high-energy physics conference held at CERN in Geneva, Switzerland, Gell-Mann held his audience captive as he outlined at the blackboard exactly what the experimenters should look for. It was detected in 1964. Nicholas Samios and no less than thirty-two collaborators recorded a track whose signature fitted that of the previously hypothetical omega-minus (Ω^-). They bombarded the 80 inch hydrogen bubble chamber at Brookhaven National Laboratory, New York, with a beam of negative kaons. The omega-minus was created by the following reaction:

$$K^- + p \longrightarrow \Omega^- + K^+ + K^\circ$$

and then decayed as follows:

$$\Omega^- \longrightarrow \Xi^- + \pi^\circ$$

The decay violates strangeness conservation as well as isotopic spin and is thus a weak interaction. The omega's lifetime of about 10^{-10} seconds allowed it to travel several centimetres in the bubble chamber before it decayed. Measurement of the energy and momenta of its decay products gives a mass of about 1672 MeV—almost exactly as predicted. Because of its large strangeness and large mass, the omega-minus is rarely created. So far it has only been seen about thirty times the world over.

High-spin resonances

More than a hundred particle resonances have now been discovered. Most of them have found a niche in an SU(3) supermultiplet.

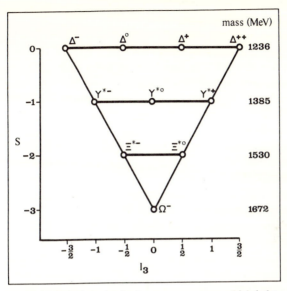

5.6. The SU(3) decuplet for baryon resonances which led to the discovery of the omega-minus hyperon

Baryons with spin 5/2 and spin 7/2 have been detected as resonances in reactions at high energies. These short-lived particles are more massive than the baryons with lower spin quantum number. Very massive mesons with spins of 2 have also been detected. The spin 5/2 baryon resonances form an octet with quantum numbers similar to those of the stable baryons (spin $\frac{1}{2}$). This similarity indicates that the more massive octet is a reoccurrence of the lighter one. In some ways there is an analogy with the electron states in an atom. Usually the electrons reside in the lowest energy configuration. If they receive extra energy, they become excited to a higher energy configuration and then decay back to the lowest energy state by emitting photons, or light. The baryon resonances with spin 5/2 may be excited states of the stable particles. During collisions the stable particles may receive sufficient energy to form new configurations which behave as short-lived particles. I would not claim that an excited atom is a new particle, and likewise resonances may not be separate particles. The question, however, is still open to debate.

Hypercharge

In the mid-1960s three Californian physicists, Geoffrey Chew, Murray Gell-Mann and Arthur Rosenfeld, proposed an alternate classification scheme which illustrated a connection between the stable particles and the resonances. In place of strangeness, they invented a new quantum number called hypercharge, with the symbol Y. Each charge multiplet has a characteristic hypercharge which is defined as twice the average charge of the multiplet; the factor of two is introduced to make hypercharge an integral number. The average charge is precisely what its name implies. For a charge triplet, such as the pions, with positive, neutral and negative members, the average charge is zero, as is the hypercharge. The nucleons have an average charge $\frac{1}{2}$ (zero for the neutron plus one for the proton divided by two); hypercharge in this case is one. For mesons, the hypercharge quantum number is identical to the strangeness quantum number, but for baryons, hypercharge equals strangeness plus baryon number. Hence hypercharge is just another way of writing strangeness. All of the strongly interacting particles fall into families which are specified by three quantum numbers—hypercharge, isotopic spin and baryon number. (There are other combinations of three quantum numbers which will do as well). Table 5.1 lists all of the ten known combinations of these three quantum numbers. Each combination corresponds to a group of particles.

Excluding the delta ($\triangle$) baryon, all of the classifications using hypercharge correspond to stable particles. Resonances also fit this classification. For example, the rho meson has the same three quantum numbers as those of the pion. The only difference between the two is their spin and mass (compare Figure 5.2 and Figure 5.4). A similar correlation exists for the rest of the strongly interacting particles. Baryon resonances can be assigned to one of the six baryon combinations listed in Table 5.1. They are more massive than their corresponding stable baryons, and they have a higher spin quantum number. Otherwise a baryon resonance has quantum numbers identical to those of a stable baryon. This correlation suggests that baryon resonances are excited states of baryons. In the final analysis, either strangeness or hypercharge will suffice as a quantum number. In practice experimenters prefer to use strangeness—largely for historical reasons: once they learn one system, there is considerable resistance to change, and strangeness was introduced prior to hypercharge. On the other hand, some theoreticians have adopted the new terminology as it is easier to manipulate in the equations. For our purposes in this book it makes no difference. Thus I have preferred the historical approach and shall use the term strangeness.

Class	Name	Symbol	Baryon Number	Hypercharge	Isotopic Spin	Multiplicity	Strangeness
Mesons	Eta	η	0	0	0	1	0
	Pion	π	0	0	1	3	0
	Kaon	K	0	+1	$\frac{1}{2}$	2	+1
	Antikaon	$\overline{K}$	0	−1	$\frac{1}{2}$	2	−1
Baryons	Lambda	Λ	1	0	0	1	−1
	Sigma	Σ	1	0	1	3	−1
	Nucleon	N	1	+1	$\frac{1}{2}$	2	0
	Xi	Ξ	1	−1	$\frac{1}{2}$	2	−2
	Omega	Ω	1	−2	0	1	−3
	Delta	$\triangle$	1	+1	3/2	4	0

Table 5.1. Families of strongly interacting particles

Quarks

The necessity of three independent quantum numbers to describe all the strongly interacting particles led Gell-Mann and, independently, George Zweig, also of the California Institute of Technology, to propose in 1964 that all the hadrons (strongly interacting particles) can be built from three ultra elementary particles. Gell-Mann named these hypothetical particles 'quarks' from a passage in James Joyce's *Finnegan's Wake*. German physicists often wince when they hear the word quark, since it means 'slime' in their native tongue.

Although quarks were originally postulated as a mathematical convenience for classifying elementary particles, some physicists believe that they actually exist. If they are discovered, quarks may be the most fundamental of elementary particles. Although there have been some disputed claims, quarks have not been detected. One participant in the quark hunt remarked that the places searched sound like a quotation from the Bible: 'they hunted them in the heavens above, in the earth beneath, and in those things in the waters and under the earth'.

Quarks should be relatively easy to detect if they do exist. According to the theory, quarks possess a fractional electric charge—either 1/3 or 2/3 of the basic charge on the electron. As the density of a track left by an ionizing particle in a bubble chamber, in a cloud chamber, or in an emulsion is proportional to the particle's charge, quarks would leave very distinctive tracks. Several groups who have hunted quarks in the heavens above thought that they saw tracks left by fractionally charged particles. Unfortunately for them, no one else could repeat these experiments. If the experimenters knew the quark's mass, their search might be facilitated. The quark's mass, however, cannot be predicted from the theory on account of uncertainties about the strength of the quark's interactions with other particles. Yet it does seem fairly certain that if it exists the quark is much more massive than any known particle.

In the SU(3) scheme, baryons can be described as a combination of three quarks, and mesons would be composed of two quarks. Let us consider the baryons and assume that all quarks have spin $\frac{1}{2}$. As baryons have half-integral spin, they must be composed of an odd number of quarks, and three is the minimum number which will

CLASSIFICATION OF PARTICLES

do. In order to construct the necessary isotopic spin states from the quarks, two of the quarks need to have non-zero isotopic spin. In analogy with the nucleons, they are sometimes called the 'p' quark and the 'n' quark, with total isotopic spin $\frac{1}{2}$ and third component equal to $+\frac{1}{2}$ and $-\frac{1}{2}$ respectively. These two quarks possess zero strangeness, but the third quark with strangeness equal to -1 is needed to form strange particles. It is sometimes called the 'λ' quark and has zero isotopic spin.

Each of these three quarks has baryon number of 1/3 so that three of them form a particle with baryon number equal to one. Now it is a simple matter to determine the charge of the three quarks. In chapter 4, I noted that the electric charge Q, the third component of isotopic spin, I_3, the baryon number B, and the strangeness S, are related by the following equation: $Q = I_3 + (B+S)/2$. Solving for Q in each case gives fractional charges for the quarks. The charges are listed in Table 5.2 together with the other quantum numbers.

Symbol	Charge	Spin	Isotopic spin		Baryon Number	Strangeness
			Total	Third Component		
'p'	$+\frac{2}{3}$	$\frac{1}{2}$	$\frac{1}{2}$	$+\frac{1}{2}$	$\frac{1}{3}$	0
'n'	$-\frac{1}{3}$	$\frac{1}{2}$	$\frac{1}{2}$	$-\frac{1}{2}$	$\frac{1}{3}$	0
'λ'	$-\frac{1}{3}$	$\frac{1}{2}$	0	0	$\frac{1}{3}$	-1

Table 5.2. Quantum properties of three fundamental quarks (see also Figure 5.7)

When these quarks are combined in every possible way, each combination of these hypothetical particles makes a different baryon in the decuplet containing the omega-minus. See Figure 5.7.

Further types of quarks are needed to explain the other supermultiplets. The meson supermultiplets are built up from quark-antiquark pairs. Antiquarks fill the same niche in the 'quark world' as known antiparticles do in the 'real world'. In order to build up the eightfold supermultiplet composed of the stable baryons, it is necessary to include other quarks with a spin projection of $-\frac{1}{2}$ as well as the first set with spin $+\frac{1}{2}$. The addition of this new set is part of a higher order symmetry than SU(3). This broader-ranging symmetry connects all of the baryon supermultiplets; it is called SU(6) and depends on six basic states. All of the meson supermultiplets can be explained by SU(6).

Many of the properties of elementary particles can be explained by assuming that quarks are the primordial particles and that they form the bases of all matter. If quarks are not discovered, would the validity of many theoretical calculations based on quarks be undermined? According to Gell-Mann, the answer is no. He invented quarks as a mathematical tool for classifying elementary particles, and at the time made no assertions that quarks existed in nature. This situation is not unusual in physics. As mathematics comprises a large part of the language of physics, it is often difficult or even impossible to explain some of the equations in physical terms—a state of affairs that can be disconcerting even to the most obscure theorist. Life among the elementary particles might be a lot simpler if quarks were detected, but many physicists would give odds that they are only a figment of Gell-Mann's imagination.

Dynamical structure

The symmetries which give rise to the supermultiplet structure of the particles do not answer all the questions we may pose about phenomena in the submicroscopic world. The symmetries tell us about the general shape of the supermultiplet and about the mass-energy difference between its members. The symmetries, however, reveal very little about the strength and nature of the forces which act on the particles. Furthermore, it is not possible to calculate the mass of any of the particles with only a knowledge of the SU(3) and

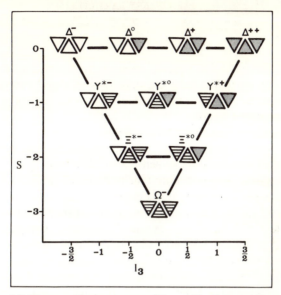

5.7. Three quarks in every possible combination recreate the baryon-resonance decuplet. See Table 5.2 for the properties of the three quarks

related symmetries. To overcome this deficiency it is necessary to know the details of how the baryons and mesons interact among themselves. Physicists call this knowledge 'the dynamics of the system'. Unfortunately no one has been able to describe adequately the dynamics of strongly interacting particles. It is one of the most important unsolved problems of particle physics. The man (or woman) who reveals the mechanism underlying the 'periodic table' of the strongly interacting particles will ensure for himself a reputation comparable to that of the greatest figures in physics.

Leptons

Things are ostensibly simpler when we try to classify the weakly interacting particles; there are only four of them. Particles which are untouched by the strong force and which participate in the weak force are called leptons. They include the electron, the muon, and their respective neutrinos. All of these particles have antiparticles as well. Unlike the baryons and mesons, no heavier leptons have been detected. There is some evidence that they exist, but it is highly controversial. Thus with only four particles to correlate, one might think that the job is easy. It is a complete mystery, however, why the muon and its neutrino exist. Every function they perform could be performed with equal ease by the electron and its neutrino. Every test that physicists have been able to devise has failed to show any difference between the electron and the muon other than their mass difference. The muon poses one of physics' gravest problems.

The breakdown of parity conservation in weak interactions has important consequences for the leptons. The particles are left-handed and the antiparticles are right-handed. In normal weak interactions the particles (electrons, negative muons and two types of neutrinos) behave as if they were left-handed screws; that is, an observer thinks that they spin clockwise when they are travelling towards him. The antiparticles (positrons, positive muons and two types of antineutrinos) behave as right-handed screws; an observer thinks they are spinning counterclockwise as they approach him.

The nature of the weak interaction, with its violation of such established symmetries as parity, charge conjugation, isotopic spin, and strangeness, is another of physics' gravest problems. At present, experiments with high-energy neutrinos offer the best hope of unravelling this mystery. Neutrinos are not affected by either strong or electromagnetic interactions. Their exclusive participation in weak interactions is a boon to the experimenters. I mentioned in chapter 2 that neutrinos can traverse the centre of the earth without interacting once, but these neutrinos are low-energy ones. As its energy increases, a neutrino has a greater probability of interacting. In the largest particle accelerators, high energy neutrinos can be created from decays of energetic pions and kaons. These neutrinos have created recognizable events in bubble chambers. As neutrino events accumulate, the structure of the weak interaction will undoubtedly become clearer.

The photon

The photon is in a class by itself. It only participates in electromagnetic interaction (I am neglecting gravitational interactions in these discussions—all particles, even the neutrino, are affected by gravity). Strong and weak interactions are not in the photon's domain of experience. As the quanta of the electromagnetic field, the photons are exchanged between charged particles when they interact. When a particle annihilates with its antiparticle the end product is often photons. The photon is its own antiparticle; under some circumstances it can disappear and create a particle-antiparticle pair, such as an electron and a positron.

The photon is not so much of a mystery as the other particles. The theory of electromagnetism is almost perfect; it is the most advanced of physical theories. The more sophisticated work started with Maxwell and Faraday in the nineteenth century. It reached its pinnacle with the twentieth-century physicists Richard Feynman (California Institute of Technology), Julian Schwinger (Harvard University) and the Japanese physicist Shinichiro Tomonaga. They received the Nobel prize for solving some long-standing mysteries about the interaction of photons and electrons. Their theories of the so-called 'quantum electrodynamics' stand as a tribute to abstract thinking of the highest order.

6
Select Problems from Modern Research

Research in particle physics is rich and varied. Some physicists are eclectic and investigate several disparate problems. Others specialize in either weak, electromagnetic or strong interactions. Ultimately they hope to develop one unifying theory which encompasses all the forces and all the particles. In the meantime, there are plenty of significant unanswered questions to keep a man occupied for a lifetime in one small area of particle physics. Occasionally within the first five chapters I discussed some unsolved enigmas of current interest to particle physicists. In this chapter I shall concentrate on two separate topics—the internal structure of nucleons, and the problem of CP violation in kaon decays. They are both the subjects of lively investigations in high-energy physics laboratories around the world.

Internal structure of the proton and of the neutron

How big is a proton? Does it have a finite extent? If so, what is its structure? Prior to the first experiments on the internal structure of the proton, physicists had a good idea that it was not a 'point' particle. One of the best theoretical reasons has to do with the existence of the proton's 'magnetic moment'. A particle has a magnetic moment if it generates a magnetic field. A simple image is that the particle acts as if it were a tiny magnet. The neutron also has a magnetic moment. As it is easier to understand how the neutron's magnetic moment leads to a finite size for nucleons, I shall consider the neutron rather than the proton. If we remember, however, that the proton and the neutron are almost the same particle, a slightly different argument should indicate that the proton has finite size.

There is an intimate connection between electricity and magnetism. Electric charge flowing in a wire will produce a magnetic field. A spinning, electrically charged sphere will produce a magnetic field similar to that produced by a bar magnet. Thus it is interesting to note that only particles with finite spin have a non-zero magnetic moment. The magnetic moment probably arises from rotation of the particle's charge. In this case, however, why should the neutron, which has no electric charge, have a magnetic moment? A possible answer is that the neutron has two parts, a positively charged one and a negatively charged one. The charges on these two parts add up to zero, but if the two parts rotated differently, their respective magnetic moments would not cancel. Consequently, the neutron would have a finite magnetic moment. The catch is that the positive and negative charges have to be spatially separated in order to rotate differently. The end result of this simple theoretical argument is that the neutron is larger than a point. Early in this century, Rutherford collected experimental evidence which supported spatially extended nucleons (protons and neutrons).

Rutherford scattering

Lord Rutherford pioneered the present-day techniques for probing nucleons when he started in 1906 to investigate the internal structure of atoms. If they had an opinion on the matter, most scientists at that time believed that atoms were homogeneous. Rutherford's brilliant experiments showed that most of the mass of an atom is concentrated in the central nucleus which is at least 100,000 times smaller than the atom's size. His experimental technique was very straightforward. He placed a radioactive source of alpha particles (the nuclei of helium atoms) in an evacuated chamber. Slits in front of the source collimated the alpha particles into a narrow beam which impinged on a target. The target was a thin metallic foil. After interacting with the target, the alpha particles were scattered at many different angles. Rutherford and his two assistants, Hans

Geiger and Ernest Marsden, measured the angular distribution of alpha particles emerging from the target. As a detector they used a screen which produced a small flash of light when struck by an alpha particle. For many hours, the three men took turns sitting in the dark and counting the number of flashes which occurred when the screen was placed at different angles relative to the alpha beam.

Rutherford and his assistants noticed that the number of particles scattered near the backward direction (that is, their direction of travel changed by almost 180 degrees) was much larger than could be expected for a homogeneous atom. See Figure 6.1. Rutherford's theoretical analysis strongly supported the existence of the nucleus. A diffuse object would not cause such large-angle scattering of the alpha particles. Originally Rutherford thought that the nucleus was concentrated at a point, but by refining his measurements, he noticed that the angular distribution of the scattered particles deviated from that expected for a point nucleus. In the year 1919 (thirteen years after starting the experiments) Rutherford showed that the approximate size of the nucleus is 10^{-12} centimetres. Yukawa later used this information to estimate the range of the strong force and to predict the mass of the pion.

The size of the proton was not determined until 1954 when Robert Hofstadter scattered high-energy electrons from a hydrogen target. He showed that the proton has a diameter of 0.74×10^{-13} centimetres. Most particle physicists were surprised that it is so large in comparison with complex nuclei. The story of why it took almost fifty years from the time of the first Rutherford experiment to the time of the first successful experiment on the proton's structure illustrates one of the most important consequences of the duality between particles and waves.

The wavelength of particles

In order to explore the interior of an object as small as a proton, the probe needs to have even smaller dimensions. The wavelength of visible light, at about 5×10^{-5} centimetres, is much too large to be of any use. Using light to probe the nucleus would be like using the waves created by a huge oil tanker to explore the size of a water bug. The reflections from the bug would have a negligible effect on the tanker's wake. Fortunately there are waves capable of mapping the

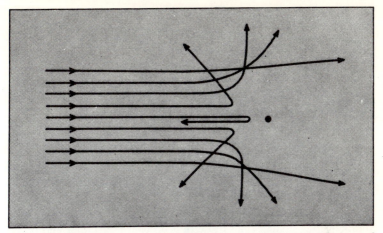

6.1. The scattering angle of one particle upon another depends upon the form of the interaction and upon the incident particle's trajectory and energy. When Rutherford detected numerous large-angle scatterings of alpha particles on nuclei, he deduced that nuclei are very compact

proton's interior; these are the waves associated with high-energy particles. In chapter 1, I briefly mentioned that particles sometimes act as waves and that they can be described by a 'wave function'. This hypothesis was drawn by Louis de Broglie in 1924. After puzzling about the nature of the photon for several years, de Broglie came to the conclusion that it must act simultaneously as a wave and as a particle. By analogy, he figured that the same duality must hold for objects generally considered as particles. He presented the essential ideas of his theory to the faculty of science at Paris University while still a student. A few years later, two Americans, Clinton Davisson and Lester Germer, demonstrated that when electrons scatter from crystals they exhibit interference and diffraction properties associated with waves.

De Broglie felt that the wavelength of a particle is just as real as the particle's mass. Furthermore, he derived a simple mathematical relationship between the particle's wavelength and its momentum, namely $\lambda = h/p$, where λ is the wavelength, p is the particle's momentum and h is Planck's constant. Planck's constant is an essential feature of quantum theory; it is related to the size of the different quanta, such as those of energy and of angular momentum. Because h is very small, we do not notice quantum phenomena on our scale of experience.

From de Broglie's simple relation, it is clear that if a particle has a large momentum, its wavelength is small. If the momentum can be made large enough, it may be possible to diminish its wavelength to dimensions smaller than those of a proton. Hofstadter and other researchers investigating proton structure used electrons as probes. As far as we know, electrons do not have an internal structure. Thus results obtained with electrons will not be complicated by any structure in addition to that of the proton. At energies of about 2000 MeV (for definition of MeV, see footnote on p.36 above) an electron's wavelength is so small that it can resolve two points separated by 10^{-14} centimetres. This distance is about one-eighth of the proton's radius. Hofstadter's original experiments were at the comparatively low electron energy of 190 MeV, corresponding to wavelengths of about one proton radius. In Rutherford's time, machines capable of accelerating particles to such high energies did not exist. It took physicists many years to develop these particle accelerators. The present largest electron accelerator in the world is at Stanford University, California. It is a two-mile-long linear accelerator capable of boosting electrons to 21,000 MeV.

The proton's electromagnetic structure

When high-energy electrons interact with a target containing many protons, they primarily interact with the electric and magnetic structure of the proton. There is no guarantee that the distribution of electric and magnetic charge within the proton coincides with those parts of the proton which take part in strong or gravitational interactions. Experiments with high-energy proton-proton interactions should in principle reveal the structure due to strong forces. Although the data are extremely complicated and difficult to analyse theoretically, it is quite clear that this structure is similar to, but not the same as, the proton's electromagnetic structure.

Hofstadter's experiments on proton structure were similar to Rutherford's experiments on atomic structure. After scattering the high-energy electrons from a hydrogen target, Hofstadter measured the angular distribution of the deflected electrons. By comparing his results with those theoretically expected for a point particle, he came to the conclusion that the proton has an electromagnetic radius of about 10^{-13} centimetres. In 1961 he received the Nobel Prize for this research. As he refined his measurements, Hofstadter's data showed that the proton is not a sphere, but has a diffuse boundary with a finite thickness.

Now that we know that the proton has a finite extension, we can inquire about what causes it to extend over space. Is it purely a rigid geometrical shape, or is it due to some sort of internal dynamics? In other words, does the proton consist of several components which are mobile in a way similar to the atom, which is composed of a nucleus and dynamic electrons? If so, then the proton could be excited to different configurations, which may correspond to baryon resonances.

Virtual particles

The shape of the proton can be explained to a crude approximation by the meson theory of nuclear forces. According to this theory, the proton spends part of its time as a neutron and a positive pion. It rapidly fluctuates between its proton state and the neutron-pion state. It is the neutron-pion state which creates the finite size of the proton detected by electronic scattering. The astute reader may note that energy is not conserved when a proton dissociates into a neutron and a pion. (The sum of the masses of the neutron and pion is greater than the proton's mass.) These two particles are not real in the sense that they cannot be detected. In the real world this event is not possible, since energy would have to be conserved. In the quantum mechanical picture of matter, however, energy conservation can be violated for a very short amount of time. This apparently heretical statement is a consequence of the Heisenberg uncertainty principle.

As I mentioned in chapter 1, a particle's position and velocity cannot be simultaneously known. Equivalently, it is possible to relate the uncertainty in the particle's measured energy, $\triangle E$, to the uncertainty of the time in which the event is measured, $\triangle t$. The product of the two uncertain quantities cannot exceed the value of Planck's constant, $\triangle E \cdot \triangle t = h$, which is an exceedingly small number. Within the bounds of the uncertainty principle, extra energy can appear in a system as long as it does so only for a very short time. In other words, new particles can appear for immeasurably brief periods. As these particles are hidden within the system and are unamenable to detection, they are called 'virtual particles'. Physicists assume that they are present because theoretical calculations agree with their existence.

We have already encountered virtual particles in chapter 3, where I mentioned that the electromagnetic force between two charged particles is carried by photons. (See Figure 3.1.) As exchanged photons can carry untenable amounts of energy and momenta, they are virtual particles. There are also real photons, but they are constrained by all the conservation laws.

As I mentioned earlier in this chapter, the magnetic moment of the neutron could be explained if it were composed of separated charged particles. Armed with the concept of virtual particles, we can surmise that the neutron dissociates into a proton and a negative pion for a short time. This simple model gives remarkable agreement with the measured magnetic moment and size of the neutron. As it is impossible to build targets of free neutrons, Hofstadter calculated the size of the neutron by subtracting the proton's size from that of the deuteron, a loosely bound nucleus composed of a proton and a neutron.

Virtual meson cloud

In his meson theory of the nuclear force, Yukawa conjectured that nucleons should be surrounded by a cloud of virtual pions extending out to 10^{-13} centimetres. Hofstadter at Stanford University, and Robert Wilson at Cornell University, New York, hoped to observe the cloud with their high-energy electrons. Their results were commensurate with a cloud of mesons, but not exactly as Yukawa predicted.

Yoichiro Nambu, a Japanese physicist working at the University of Chicago, showed in 1957 that the meson cloud could be explained by the presence of meson resonances with one unit of spin. These resonances can decay into pions. In previous chapters we encountered the rho (ρ) meson which is a resonant state of two pions and the omega (ω) meson, which is a resonant state of three pions. Both of these mesons have one unit of spin. There are additional meson resonances which satisfy Nambu's conditions. With the discovery of these meson resonances, part of the mystery of the proton's meson cloud was cleared up.

According to this picture, when an electron scatters from a proton it exchanges a virtual photon, which couples to the proton in several different ways. The total interaction is the sum of these different couplings. As shown in Figure 6.2, the proton can be represented as a 'blob' which can be separated into several components. When the electron exchanges a photon with the blob, as a first approximation the proton acts as if it were a point particle. The next most important contributions to the scattering are a two-pion resonance (rho) and a three-pion resonance (omega). There is no good reason why even more complicated contributions to the total scattering cannot be present. For example, the photon could couple to a baryon-antibaryon pair, or even strange particles could participate. All of these combinations will come into the picture but their strengths, characterized by the mass and coupling of the virtual particles involved, will be less than those of the first three terms in Figure 6.2.

One American theoretical physicist summed up the situation quite aptly when he wrote, 'From this point of view, the "inside" of a nucleon is a complex of complicated configurations of various mesons, meson complexes, baryon pairs, baryon resonances, etc., which come into (virtual) existence and disappear again as a result of quantum fluctuations; each makes some contribution, large or small, to the charge and current density of the object we call a nucleon. This is far from the relatively simple picture of the structure of the atom, where the charge and current distribution is associated with permanent essentially stable constituents—electrons, protons, and neutrons.'

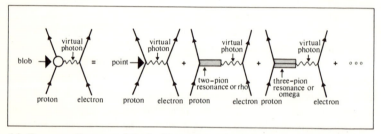

6.2. Feynman diagrams of the leading terms which contribute to electron-proton scattering. Additional terms will include baryon-antibaryon pairs and strange particles as well as other particle combinations

Repulsive core

Before moving on to the more recent studies of the nucleon's interior, I would like to mention the role played by the rho and the omega in nuclear forces. The range of the strong force is quite small—about 10^{-13} centimetres—but the force is very powerful. If the force were carried entirely by virtual pions, neighbouring nucleons in a nucleus would be pulled together and collapse into each other as a particle-antiparticle pair does. However, it is a property of the spin-one particles, such as the rho, that they transmit a repulsive force between identical particles. Rho mesons within the central core of nucleons prevent them from collapsing. Figure 6.3 illustrates how the energy well of one nucleon might look to another nucleon. The barrier near the centre represents the 'vector mesons' of which the rho meson is one.

Partons and bootstraps

The most recent research at the Stanford linear electron accelerator refutes this relatively simple picture of the nucleon dominated by two-pion and three-pion resonances. Electron scattering at the highest energies available is reminiscent of Rutherford's original experiments. The tentative conclusion of the data analysis is that nucleons are composed of point-like entities now called partons. Furthermore these partons seem to have something in common with the hypothetical quarks discussed in chapter 5. These results are very different from what anyone had predicted. They are still speculative, but if these experiments are substantiated, physicists may be back on the trail of a particle hierarchy with partons (or quarks) on the bottom and everything else composed of them.

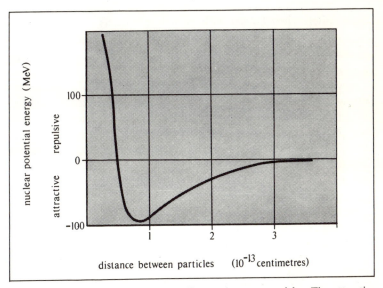

6.3. The nuclear force plotted as the distance between particles. The attractive part of the force is short range extending over a limited distance. The repulsive core is due to the presence of the rho meson and other vector mesons within protons and neutrons

The alternative model of particles is more democratic. As we saw in the case of nucleon structure, all the elementary particles are busy disassociating and changing into one another. The proton and the neutron are not outstanding because they are more fundamental than the other particles, but because they have been more accessible to experimental observation. A system where each strongly interacting particle helps to generate other particles, which in turn help to generate the original particle, was developed by Geoffrey Chew, University of California. He called his model 'bootstrap' dynamics because each particle 'holds itself up by its bootstraps'. Whether you favour a hierarchical or democratic model of particles may be a matter of personal taste, as is so often the case with practising physicists. This question, however, will most likely remain unresolved for many years. The solution, when it does come, will probably arise from a most unlikely source and will undoubtedly surprise everyone.

CP violation in kaon decays

Quite often when I meet with particle physicists I ask them what they think is the most interesting problem facing them. The majority of them answer that CP violation in kaon decay poses the biggest problems and is the most exciting line of research. As I mentioned in chapter 3, CP violation strikes at the heart of one of physics' most important symmetries, CPT symmetry, where C is charge conjugation (the conversion of particles into antiparticles and vice-versa; see chapter 4), P is parity (right-left symmetry; see chapters 3 and 4) and T is time reversal. If we are to believe the basis of modern physics, then CPT is conserved in every particle reaction. This means that if, in a particle event, every particle is converted into its corresponding antiparticle (and vice-versa), and the reaction is reflected through a mirror, and the axis of time is reversed, then the resulting event must be as real as the original one. This is called CPT conservation. It can be derived from the assumption that particle physics is governed by the laws of quantum mechanics and by the theory of relativity. If CPT falls, so does the framework of modern physics—a disastrous state of affairs for physicists.

In 1964, four physicists from Princeton University, James Christenson, James Cronin, Val Fitch and René Turlay, published evidence that the combined symmetry CP is violated in the decays of neutral kaons. Since then other experiments have confirmed this result. If CP is violated, then T must be violated by a compensating amount, if CPT is to remain intact. Time reversal violation causes some theoretical problems, but they are not as severe as the ones caused by CPT violation. There are some important theorems in particle physics which depend on the complete symmetry of time reversal, and it had always been assumed that T was a valid symmetry. Numerous experiments to check for T violation in weak, strong and electromagnetic interactions have shown no violation. Thus CPT hung in the balance. Most recently, theoretical physicists have proposed a new force, the superweak (see chapter 3), which absorbs all of the problems. None the less, the experiments still continue as many physicists are not satisfied. The main result so far is that the neutral kaons are very unusual particles.

Properties of kaons

Kaons, or K mesons, are unique among the stable particles. They have a neutral particle, $K°$, with a distinct neutral antiparticle, $\overline{K°}$. These particles are created in association with other strange particles. When they decay, both the $K°$ and $\overline{K°}$ can decay into either two pions or three pions. By studying the two and three pion decay modes of charged kaons, Lee and Yang first proposed parity violation in weak interactions. The neutral kaons, however, are more unusual. The two decay processes have greatly different lifetimes. The two-pion decay occurs relatively rapidly, in about 10^{-10} seconds. It can be written:

$$K° \longrightarrow \pi^+ + \pi^-$$
$$\overline{K°} \longrightarrow \pi^+ + \pi^-$$
or
$$K° \longrightarrow \pi° + \pi°$$
$$\overline{K°} \longrightarrow \pi° + \pi°$$

The three pion decay is longer, about 4×10^{-8} seconds. It can be written:

$$K° \longrightarrow \pi^+ + \pi^- + \pi°$$
$$\overline{K°} \longrightarrow \pi^+ + \pi^- + \pi°$$

This disparity in the lifetime was difficult to explain until Murray Gell-Mann and A. Pais showed that it is a consequence of the wave nature of particles. They proposed that the $K°$ and $\overline{K}°$ can combine like waves to produce two other particles called $K_1°$ and $K_2°$. In the first combination, $K_1°$, the probability for decay into two pions is enhanced while in the second combination, $K_2°$, it is cancelled. Both of the strongly interacting, strange particles $K°$ and $\overline{K}°$ have components of $K_1°$ and $K_2°$ in them and will decay fifty per cent of the time in each mode. The rapid decay into two pions is via $K_1°$ and the slower three pion decay is via $K_2°$. In a beam of neutral kaons, the short-lived component, $K_1°$, decays away, leaving the longer lived one, $K_2°$, which decays by three pions. Fitch and his colleagues at Princeton noticed that $K_2°$ can also decay into two pions. Although a rare event (it occurs only once in every 500 three-pion decays), the two pion decay mode of the long-lived neutral kaon provided the evidence that CP is violated.

Consider the two pion decay mode of a neutral kaon. As Figure 6.4 shows, it has an even (+) value for the CP symmetry. Charge conjugation converts the negative pion to its positive counterpart, and the positive pion changes into a negative one. The parity reflection, through the position formerly occupied by the kaon, puts everything back as it was. With three pions, however, things are different. Pions have odd intrinsic parity (see chapter 4); a combination of three of them can have odd parity. Thus they can also have odd (−) value for the CP symmetry. The superposition of $K°$ and $\overline{K}°$ which forms $K_2°$ has an antisymmetric wave function with respect to CP; thus the long-lived neutral kaon would always decay into three pions if CP were conserved. Its decay into two pions shows that CP is a broken symmetry.

Preserving CPT

Attempting to explain the CP violation of neutral kaon decays, physicists started to question all of their assumptions about discrete symmetries. They looked for CP violation in other systems of particles. As CP violation cast aspersions on T conservation and on CPT conservation, they looked for violations of these symmetries in weak, strong and electromagnetic interactions. So far, no violation has been found. They also searched for the separate breakdown of

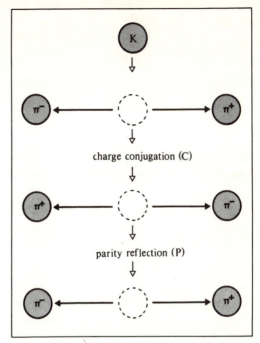

6.4. The two pion decay of neutral kaons returns to its original configuration after the combined operations of charge conjugation (C) and parity reflection (P). This means that CP symmetry is even for this decay. The parity reflection is about the position where the kaon disappeared

C and P in electromagnetic and strong forces (we already know that weak interactions violate these symmetries—see chapter 3). All of the evidence seems to support the existence of a new force, the superweak, which violates both CP and T, but leaves CPT intact. Physicists have not had enough time to examine the implications of the breakdown of T in this new force. Since CP violation is so rare, however, T violation cannot have much of an effect on the foundations of particle physics. The superweak force has not been manifest in any system except neutral kaons. Yet time reversal is no longer a hallowed symmetry.

Eta decay

One interesting spin-off from experiments relating to CP violation is the search for C violation in electromagnetic interactions. The eta (η) meson is a good experimental system as it decays via the electromagnetic force into three pions:

$$\eta^\circ \longrightarrow \pi^+ + \pi^- + \pi^\circ$$

To test for charge conjugation violation, the experimenters compare the energies of the charged pions. If the negative and positive pions have the same energy of motion, on the average, then C is a valid symmetry. However, if, for example, the positive pion carries away more energy, the situation would be different in the antimatter world where the negative pion would possess more energy. The experimental situation has fluctuated between C violation and C conservation with the experiments improving all the time. The final word on the subject is yet to be written.

If C is violated in eta decay, it will have fascinating consequences. We would, in theory, be able to determine if distant parts of the universe are composed of antiparticles. This experiment would get over the pitfalls in the cobalt-60 parity violating experiment which prevented us from distinguishing matter from antimatter (see chapter 3). If we find that positive pions from eta decay are, in fact, more energetic on the average than negative pions, this situation should hold in a distant galaxy. All we need to do is contact intelligent life on planets in those galaxies and ask them to do the same experiment. If the sign of the electric charge on the more energetic pion is the same as that on the nuclei of their atoms, we should encourage closer contact. If, however, they are of opposite signs, we should avoid that part of the universe or risk being annihilated.

Epilogue

For hundreds of years natural philosophers have been searching for the essence of matter. In his seventeenth-century treatise, '*Opticks*', Sir Isaac Newton stated their case with these timeless remarks:

'Now the smallest Particles of Matter may cohere by the strongest Attractions, and compose bigger Particles of weaker Virtue; and many of these may cohere and compose bigger Particles whose Virtue is still weaker, and so on for divers Successions, until the Progression end in the biggest Particles on which the Operations of Chymistry, and the Colours of natural Bodies depend, and which by cohering compose Bodies of a sensible magnitude.

'There are therefore Agents in Nature able to make the Particles of Bodies stick together by very strong Attractions. And it is the Business of experimental Philosophy to find them out'.

Newton's sentiments were echoed in the twentieth century by Robert Hofstadter, who concluded his Nobel speech with the following words:

'One can only guess at future problems and future progress, but my personal conviction is that the search for ever-smaller and ever-more-fundamental particles will go on as long as man retains the curiosity he has always demonstrated.'

Appendix
Suggestions for Further Reading

Books containing semipopular accounts of particle physics

Particle Physics: The High-Energy Frontier; M. Stanley Livingstone (McGraw-Hill, 1968).
The Nuclear Apple; P. T. Matthews (Chatto & Windus, 1971).
Elementary Particles; David H. Frisch and Alan M. Thorndike (D. Van Nostrand Co. Inc., 1964).
The Fundamental Particles; C. E. Swartz (Addison-Wesley, 1965).
The World of Elementary Particles; K. W. Ford (Blaisdell, 1963).
Tracking Down Particles; R. D. Hill (Benjamin, New York, 1964).

Articles in the monthly *Scientific American*

'The Structure of the Proton and the Neutron'; Henry Kendall and Wolfgang Panofsky, June 1971.
'Photons as Hadrons'; Frederick Murphy and David Yount, July 1971.
'Can time go backward?'; Martin Gardner, Jan 1967.
'Violations of Symmetry in physics'; Eugene Wigner, Dec 1965.
'The Omega-Minus Experiment'; W. B. Fowler and N. P. Samios, Oct 1964.
'Strongly Interacting Particles'; Geoffrey Chew, Murray Gell-Mann and Arthur Rosenfeld, Feb 1964.

'Resonance Particles'; R. D. Hill, Jan 1963.
'Conservation Laws'; G. Feinberg and M. Goldhaber, Oct 1963.
'The Spark Chamber'; Gerard K. O'Neill, Aug 1962.
'Particle Accelerators'; Robert R. Wilson, March 1958.
'The Bubble Chamber'; Donald A. Glaser, Feb 1955.
(Many articles from *Scientific American* are separately available as *Scientific American Offprints*.)

Excellent semipopular articles on particle physics appear almost weekly in the British journal *New Scientist*.
More advanced articles, which are still comprehensible to non-physicists appear in the following journals:
Science (USA), *Physics Today* (USA), *CERN Courier* (Switzerland).

Elementary books using mathematics

Nuclei and Particles; Emilio Segré (Benjamin, New York, 1964).
Quantum Physics; Eyvind H. Wichmann (McGraw-Hill, 1967, 1971).
The World of the Atom; edited by Henry A. Boorse and Lloyd Motz (Basic Books Inc., 1966). [I highly recommend this two-volume book. It contains interesting historical and biographical material, including many Nobel prize speeches, as well as clear presentations of the pertinent physics.]

Glossary

accelerator: a machine designed to produce high-energy particles.

antimatter: particles which annihilate both themselves and their complement in ordinary matter when they come into mutual contact.

baryon: the class of strongly interacting particles with half-integral spin. It includes nucleons and hyperons.

beta decay: the process whereby an atom of one element transforms into an atom of a different element by emitting an electron from its atomic nucleus.

bubble chamber: a device which records the tracks of charged particles which pass through it. It is usually filled with a low-temperature liquid gas, such as liquid hydrogen, which forms bubbles along the particle's path.

charge multiplet: a family of particles with identical properties except for their electric charge.

collimate: to line up several small openings cut in metal so that a beam of particles passing through them will have a well-defined spatial position and direction of travel.

cosmic rays: high-energy particles impinging on the earth from an unknown source in outer space.

deuteron: the nucleus of heavy hydrogen. It consists of one proton bound to one neutron.

emulsion: photographic film used to record the tracks of charged particles which penetrate through it.

gamma rays: high-energy electromagnetic radiation similar in character to X-rays, ultraviolet radiation, visible light, infrared radiation and radio waves.

hadron: a name for all particles which experience the strong nuclear force.

helicity: a relation between the directions of a particle's rotation and its velocity. A particle with positive helicity behaves similar to a right-hand screw, while a particle with negative helicity behaves similar to a left-hand screw.

hyperon: a class of strongly interacting particles which are more massive than protons or neutrons and which have half-integral spin.

ion: an atom with more or less than the full complement of electrons necessary to make it electrically neutral.

isotopic spin: a quantum number used to specify the number of particles in a charge multiplet and to specify the electric charge on each particle.

lepton: the class of light-weight particles which do not experience the strong nuclear force.

magnetic moment: a property of rotating particles which gives them a small magnetic field similar to that of a bar magnet.

meson: the class of strongly interacting particles with integral spin and with mass less than that of the proton and the neutron.

nucleon: a family name for the proton and the neutron.

parity: right-left or mirror symmetry of particles and their interactions. According to the notion of parity conservation, the mirror image of a particle reaction conforms to all of the laws of physics.

photon: the particle-like quantum of the electromagnetic force.

propagator: a particle which transmits a force between two other particles.

quantum: a discrete unit or packet of a quantity. At the atomic level some quantities, such as energy, can only be exchanged between two particles in quantum amounts.

GLOSSARY

range: the distance over which a force is effective.

resonance: extremely short-lived particles or excited configurations of longer-lived particles.

scattering: the interaction of two particles in which they either rebound off each other similar to billiard balls or create new particles.

spin: the quantum number which characterizes the magnitude of a particle's rotation. In units of Planck's constant, a particle's spin is restricted by the quantum theory to either integral or half-integral values.

supermultiplet: a family of particles which can be grouped together using SU(3) symmetry.

wave function: a mathematical function which has wave properties and which describes a particle.

Index

Accelerator, 34, 35–36
Alpha particles, 21, 23, 101–102
Alvarez, Luis, and bubble chambers, 37; and particle resonances, 88, 90
Anderson, C. D., and mesons, 28; and the positron, 25–27
Angular momentum; conservation, 63–66; definition, 63; see spin
Antimatter, 24–26, 76
Antiproton, 67, 71, 85; discovery, 42
Associated production, 76–77
Atomic concept, 10
Atoms, 9–11, 13, 14, 17, 20, 43, 64, 71, 91
Baryon, 92; classification, 81–93; conservation, 66–68; relation to quarks, 95–96; resonances, 88–90, 92; SU(3) supermultiplets, 82; table of properties, 40–41; see hyperon and nucleon
Beta decay, 21, 24, 32, 65; in parity violation, 51–54
Bohr, Niels, and quantum theory, 16; and the atom, 20
Bootstrap model, 109–110
Bubble chamber, 37, 62
Butler, C. C., and strange particles, 34, 76
Chadwick, James, and the neutron, 23–24
Chamberlain, Owen, and the antiproton, 42
Charge; baryon, 66–68; electric, 66; hypercharge, 92–93; lepton, 68
Charge conjugation, 76–78; 126–127
Charge multiplets, 81–82

Chew, Geoffrey, and hypercharge, 92; and the bootstrap model, 110
Cloud chamber, 26
Cobalt-60 experiment, 51–52
Cosmic rays, 12, 25, 26, 28–30, 34
Cowan, R. D., and the neutrino, 32–33
CPT symmetry, 56, 110–113
CP symmetry, 72; violation, 56–57, 110–114
Cronin, James, and CP violation, 110
Cyclotron, 35
De Broglie, Louis, 80; and particle-waves, 103–104
Democritus and atoms, 10, 11
Deuteron, 70, 106
Dirac, P. A. M., and antimatter, 24–25
Eightfold way, 81
Einstein's mass-energy formula, 14, 32, 36f
Electromagnetic force, 43–49; see force
Electron, 9, 11, 13, 14, 17, 24, 25, 30, 34, 44, 46, 47, 60, 64, 66, 68, 71, 72, 98; discovery, 20; in beta decay, 21–22, 32, 51–54; proton scattering, 104–105, 109; negative-energy electrons, 25–26; see lepton
Electron volt, definition, 36f
Emulsion, 26
Energy conservation, 58
Epicurus and atoms, 11
Eta meson, 39, 85–86; and charge conjugation, 114; see meson
Fermi, Enrico, and baryon resonances, 88; and beta decay, 32; and neutrinos, 32

Feynman, Richard, and photons, 99
Feynman, diagrams, 31, 48, 108
Fitch, Val, and CP violation, 110
Force and partial symmetries, 68–79;
propagator, 47–49; range, 46–47;
repulsive core of nuclear forces, 108;
strength, 45; superstrong, 57;
superweak, 56–57
Gamma rays, 21, 24, 25, 26, 28;
see photon
Gell-Mann, Murray, and hypercharge,
92; and quarks, 57, 94; and
strangeness, 77; and SU(3), 81–82,
89–90
Glaser, Donald, and the bubble
chamber, 37
Gravitational force, 44–49; see force
Graviton, 49
Hadron, 45
Heisenberg, Werner, 80; and isotopic
spin, 72; and quantum theory, 17
Heisenberg uncertainty principle, 18,
63, 86, 106
Helicity, 54, 55
Hofstadter, Robert, 115; and the size
of the proton, 102, 104–105
Hypercharge, 92–93
Hyperon, 39, 49, 67; definition, 34–35;
table of properties, 40–41; see baryon
Ion, 21, 26, 27, 37, 60
Ionization, 24, 26, 37
Isotopic spin, 72–76; violation, 75
Kemmer, N., and the pion, 30
K mesons (kaons), 14, 51, 56–57, 67,
77, 78, 90; discovery, 39; and CP
violation, 111–112; and theta-tau
puzzle, 51; see meson and strange
particles
Lambda hyperon, 14, 34, 44, 59–60,
75, 77; see baryon, hyperon and
strange particles
Lawrence, E. O., and the cyclotron, 35
Lee, T. D., and parity, 49
Lepton, 45, 98; table of properties,
40–41
Lepton conservation, 68
Lucretius and atoms, 11
Magnetic moment, 100–101, 106
Meson, 49, 64, 67; cloud, 106–107;
history, 28–31; resonances, 85–88;
SU(3) supermultiplets, 84; table of
properties, 40–41

MeV (million electron volts)
definition, 36f
Molecules, 9–11, 14, 43
Momentum conservation, 59
Muon, 30, 33, 34, 44, 47, 55, 68, 98;
discovery, 28–30; see lepton
McMillan, E. M., and particle
accelerators, 35
Nambu, Yoichiro, and the structure
of the proton, 107
Ne'eman, Yuval, and SU(3), 81–82, 90
Neutrino, 13, 34, 44, 47, 56; discovery,
31–33; electron neutrino, 33; muon
neutrino, 33; and parity violation,
52–55; spin, 65; see lepton
Neutron, 9, 11–13, 28, 30, 44, 45, 47,
62–63, 67, 69–70, 72; discovery,
23–24; decay, 33, 65; spin, 64; see
baryon and nucleon
Newton, Isaac, 115
Nishijima, K., and strangeness, 77
Nucleon, 28, 35, 43, 49, 64, 67, 74, 81;
internal structure of, 100–110;
table of properties, 40–41; see
neutron and proton
Nucleus, 11, 14, 20, 26, 28, 43, 51–52, 64
Omega meson, 85–86, 107
Omega-minus hyperon, 88–90
Oppenheimer, J. Robert, and the
pion, 30
Pais, A., and strange particles, 76
Parity, 49–56, 69–71; intrinsic, 49–50;
violation, 49–56
Particle; definition, 12–16; mass
determination, 59–60; size, 102;
stability, 13; structure, 14; tracks,
26–27, 59–60; table of properties of
stable particles, 40–41
Particle resonances, 60–63
Parton, 109
Pauli, Wolfgang, and neutrinos, 31
Pauli exclusion principle, 70–71
Photon, 28, 34, 44, 49, 75, 99; as a
propagator, 47; table of properties,
40—41; see gamma rays
Pion (pi meson), 14, 34, 36, 39, 43, 44,
51, 59–60, 62–63, 65–66, 67, 82;
discovery, 28–31; decay, 55–56, 75;
as a propagator, 47; intrinsic parity,
69–71; isotopic spin, 74; proton
scattering, 18–19, 45, 62, 69, 71,
74–77, 88; see meson

INDEX

Planck's constant, 104
Positron, 33, 34, 46, 55, 60; discovery, 25–27; see lepton
Powell, C. F., and mesons, 29
Proton, 9, 11–14, 21–24, 28, 30, 33, 34, 39, 46, 47, 59–60, 63, 64, 66, 67, 72, 85; discovery, 21; pion scattering, see pion; see nucleon and baryon
Quantum mechanics, 16–19, 22, 80
Quark, 16, 57, 94–96, 109
Radioactivity, 21
Reines, F., and the neutrino, 32–33
Resonances, 60–63; SU(3) classification, 85–93
Rho meson, 62–66, 85, 107–108
Rochester, G. D., and strange particles, 34, 76
Rutherford, Ernest Lord, 23, 28; and the atom, 20
Rutherford-Bohr atom, 20, 80
Rutherford scattering, 101–102
Schwinger, Julian, and the photon, 99
Scintillation detectors, 37
Segré, Emilio, and the antiproton, 42
Sigma hyperon, 34, 75, 78, 82; see baryon and hyperon
Spark chamber, 37
Spin, 22, 52, 63–66, 72; see angular momentum
Strange particles, 34, 39, 76–79; resonances, 89
Strangeness, 76–78; see hypercharge

Strong nuclear force, 43–49, 108; see force
Superstrong force, 57
Superweak force, 56–57, 112–114
SU(3) symmetry, 81–96; supermultiplets, 82–90
Synchrocyclotron, 36
Synchrotron, 36
Theta-tau puzzle, 51
Thomson, J. J., and the electron, 20; and the proton, 21
Time reversal, 56–57, 112–114
Tomonaga, Schinichiro, and the photon, 99
Veksler, V., and particle accelerators, 35
V particle, 34, 76
Virtual particles, 44, 105–107
Wave function, 17, 49, 69, 103
Wave-particle duality, 17, 102–104
Weak nuclear force, 43–56; see force
Wilson, C. T. R., and the cloud chamber, 26
Wilson, Robert, and the structure of the proton, 106
W particle, 47–49
Wu, C. S., and parity, 51
Xi particle, 82; see baryon and hyperon
Yang, C. N., and parity, 49
Yukawa, Hideki, and the pion, 28, 47, 102, 106
Zweig, George, and the quark, 94